Using Paired Constraints to Solve The Innovation Problem

Patricia D. Stokes · Michael Gibbert

Using Paired Constraints to Solve The Innovation Problem

Springer

Patricia D. Stokes
Department of Psychology
Barnard College, Columbia University
New York, NY, USA

Michael Gibbert
Faculty of Communication Sciences
Universita della Svizzera Italiana
Lugano, Ticino, Switzerland

ISBN 978-3-030-25770-5 ISBN 978-3-030-25771-2 (eBook)
https://doi.org/10.1007/978-3-030-25771-2

This Edizioni della Normale imprint is published by the registered company Springer Nature Switzerland AG
The registered company address is: Gewerbestrasse 11, 6330 Cham, Switzerland

To my husband, Ron Romano, the love of my life.
—Patricia D. Stokes

To Alexander Johannes and Beatrice Theresa, my two cool kids.
—Michael Gibbert

Prefaces/Previews

Pat's Preface

Reversing the order of the title (*Using Constraints to Solve Innovation Problems*), I'll start with the two parts of an innovation problem and then briefly (very briefly) introduce a practical problem-solving framework you can use to solve both.

The Innovation Problem

The innovation problem has two parts: how do you *start*, and how do you *sustain*, innovation. The problem-solving *framework* has three parts: a *structure* (the problem space), a *strategy* (the paired constraints), and a *process* (solution-by-substitution).

Here's my short, much over-simplified, preview.

How do you start doing something new? You start very specifically. You identify a current product/style/situation to work against. This becomes the initial state in your problem space. You then select one element, just one, in that product/style/situation to preclude. Next you select/promote a substitute. Once the preclude-promote pairing begins, it becomes self-sustaining: one substitution suggests or requires another. I call the process *solution-by-substitution*. The substitution series *is* the solution path that solves the innovation problem (Stokes, 2006).

Haven't you also said that the solution path is itself the innovation?

Yes, it is. Thanks for reminding me.

How do you continue doing something new? You start over. Only now, your initial solution becomes the initial state.

Where do the substitutions come from? The *tool box*, the one in your head, the one you can't think outside of. The contents of the tool box are your *expertise*—what you know about your domain and what you can do with what you know.

These are what I call *basics*. *Borrowing* from other experts, and from other domains, are what make your tool box bigger.

Pat, you should tell them more about the tool box.

I will, later. (See Digression II).

Two Examples

Example I: Two authors, two-voices. The hardest part of writing a book is starting it. This is because all books (like all problems) are only realized, structured, and restructured as they are written.[1] The solution path for any book *is* the finished book. This book had an additional difficulty. It was not just that there were co-authors, Michael and me. It was also that we wanted the book to reflect the conversations that created it. To do this would require a number of substitutions, as shown in Table 1.

Table 1 Two-voices problem

Initial state:	Co-authors, one style		
Search space		**Constraint pairs**	
	Preclude	➔	**Promote**
	Continuous narrative		Conversation
	Single voice/style		Dual voices/styles
	Shared type face		Distinctive type faces
Goal state:	Co-authors, two styles		

The first substitution precluded a continuous narrative and, in its place, substituted a conversation, its exact form unspecified at first, but suggesting separately written sections interrupted (as conversations are) with comments, suggestions, digressions, questions. The second followed from the first: preclude a single voice and promote dual, distinct voices. The third followed from the second: distinct voices suggested distinct type faces. This is my (Pat's) type face. Michael's is *this* (*italic*).

An important, albeit partial, borrowing for the structure was a book in the form of a continuous conversation.

[1]Annie Dillard (wisely) wrote that the early chapters, especially the earliest which has become so that it is so familiar that it feels indispensable, must be revised because the book does not find its form so fast (Dillard, 1989).

Between?

Edward Said, the literary critic, and Daniel Barenboim, the pianist/conductor (Barenboim & Said, 2004).

Example II: Monet's Innovation—Impressionism. I thought we should have a more complete (albeit abbreviated) example (Stokes, 2001, 2011). So, Monet, a painter I hope is familiar to most of you. Monet's innovation, Impressionism, began with a borrowing, a color wheel published in a Parisian newspaper by a chemist named Chevreul. The wheel, which broke light up into 72 segments gave Monet both his continuing goal (*show how light breaks up*) and his first sub-goal (*on things*). Showing how depended on his existing and *expanding* expertise as a painter. The italics are important: substitutions are additions—new ways of making and of noticing—to an innovator's tool box.

Given the new goal, how did Monet *start* to do his something new?

With a single substitution (the first constraint pair in Table 2). He precluded contrasts in value (dark-light) and substituted/promoted contrasts in hue. If we think of hue as pure color, this pairing also substituted pure for mixed colors. It also suggested, in fact required, a second pairing (preclude depth, promote surface) and a third (preclude continuous brush strokes, promote a mosaic of separate color patches). The first three produce the fourth, preclude depth, and promote surface.

Table 2 Monet's first substitution series

Initial state 1:	Show how things look (traditional landscape painting)		
Search space		**Constraint pairs**	
	Preclude	➔	**Promote**
	Contrasts in value		Contrasts in hue
	Mixed colors		Pure colors
	Continuity strokes		Separate strokes
	Depth		Surface
Goal state 1:	Show how light breaks up on things		

Stop. I've seen your Monet sketches. Can't we include at least one here.

Good idea.

My cartoon version of *Regatta at Argenteuil* (Fig. 1) shows light breaking up into clear, bright, and clearly separated oblongs of pure color. The houses are sketched in separate strokes of pure cadmiums (red and orange). Their reflections overlap but do not mix with the pure blues of the sky and the pure greens of the trees and grasses. The sails and their reflections also break up (into cream-colored lozenges). The sails are closer to us, so their reflections are larger than those of house and tree, but all three (houses, trees, sailboats) sit on the same surface. Since dark-light contrasts (in value) are precluded, there is no depth separating them.

Fig. 1 Regatta at Argenteuil (1872)

Amazing but only the start—which leads to our next question: How did Monet *continue* doing something new? The answer is by *starting over*, re-structuring his own solution, changing his sub-goal from light breaking up *on* things to light breaking up *between* things. This shift led to what I call the second substitution series, summarized in Table 3.

Between things means (in French) the *enveloppe* or (in English) the atmosphere. Since the *enveloppe* is continuous, the first substitution precluded local color (the green of leaves, the blue of the sky) and promoted shared colors. Since the colors were shared, the original mosaic of separate strokes was precluded and, in its place, a continuous web of common colors was promoted. Finally, since the *enveloppe* changes continuously, the single study was precluded, and in its place, each image (poplars, haystacks, the façade of the cathedral at Rouen) was painted (again and again) on separate canvases that became a series.

Table 3 Monet's second substitution series

Initial state 2:	Show how light breaks up *on* things		
Search space		**Constraint pairs**	
	Preclude	➔	**Promote**
	Local color		Shared color
	Separate strokes		Web of shared strokes
	Single study		Series of studies
Goal state 2:	Show how light breaks up *between* things		

Images of the series are easy to find online. Just type in *Poplars*, there are over 20 paintings in the series. You should also look at images of his third and most radical innovation, the *Grandes Decorations*, which I've discussed in great detail elsewhere. Start with *Water Lilies—Reflections of the Willow.*

Michael's Preface

My interest in constraints started as a Bachelor Honors student at the University of Stellenbosch, South Africa, back in 1998. I was renting a granny flat with a parish priest. The priest was a thoughtful, bookish man and he encouraged me to explore his extensive library. One cold winter evening, looking for something to keep me in front of the one source of heat in the house (an open fireplace), I stumbled over an A-4 sized manuscript, which seemed to be a bound thesis of some kind. It wasn't thick, maybe 150 pages or so, and had the intriguing title "Modeling Complexity." Its author was the late Paul Cilliers, and it was his Ph.D. thesis at Cambridge (under the supervision of Marie Hesse).

It was, I think, the first Ph.D. thesis I ever read, and I was awestruck by some of the statements, in particular the idea that boundaries and constraints are both enabling and constraining. In the sense that while they separate, they also unify since they constitute that which they bound.

My landlord could see my fascination with the manuscript and told me that Paul Cilliers was a Professor of Philosophy at Stellenbosch University, and so we got together. The various excellent dinners at Paul's house that winter (aided by venerable old reds from the Western Cape) were foundational in my interest in the ambivalent nature of constraints, *in their role as innovation drivers. Several special issues, books, and articles later, Pat and I crossed paths and realized we had been working on the same topic, but from different directions, and the idea for this book was born.*

My background (aside from cooking) is business management, with master's and Ph.D. theses on strategic renewal. The first academic job I got was as a Professor of Marketing at Milan's Bocconi University, and the title stuck. Much of my research focused on innovation management under constraints, but constraints defined in a more traditional, standard dictionary way. This way. According to Webster's, constraint is derived from two Latin words, con *(together) and* stringere *(to draw tight). "To draw tightly together" becomes, in the standard definition, confinement or restriction, compulsion or coercion. Constraints of this kind generate a kind of innovation problem, commonly known as a problem of necessity* (Hoegel, Gibbert, & Mazursky, 2008). *Popular parlance has it that necessity is the mother of innovation, but innovation is not an immaculate conception, which leads to my first question.*

Michael's Questions

If necessity is the mother of innovation, who is the father? *I find this analogy striking. Think about it. Unless there is some kind of added ingredient, constraints are simply counterproductive when it comes to innovation. The thesis of this book is obviously that they are not counterproductive. A promising candidate for the added ingredient is the second of the constraint pairs that Pat introduced. The first, as expected, precludes. The second, unexpectedly, promotes the substitutions that father the innovation. Interestingly, very interestingly, this dual role also appears in postmodern philosophy. So, to answer Pat's earlier question, let me elaborate (briefly here) on boundaries and constraints in this more inclusive sense.*

Wait! Wait! Couldn't we extend the analogy further? The Mother of Invention (Necessity), The Father (Strategy or Philosophy), The Fairy Godmother (Opportunity).

Wait Pat. Let me follow my train of thought first.

How can boundaries both enable and constrain? *The inclusiveness here comes from seeing boundaries and constraints as both enabling and constraining. When I say "boundaries" I really mean constraints, things that hold us back. Our natural reactions to things that hold us back are either (1) to literally and figuratively break through them to get to the other side somehow or (2) to accept them as impenetrable and submit to their dividing our space. But there is another reaction, I think. And that reaction is what this book is about.*

Let's go back to Paul Cilliers' thesis (1998). *Boundaries separate one thing from another and yet, automatically, they are also part of the things that they set apart. They are both the problem and the solution. Many approaches in systems thinking rely on a similar idea—the interdependence of problem and solution. For example,* obliquity *refers, not to taking a direct solution path (the one which perhaps is most immediately available, though not necessarily the most efficient or creative), but rather to taking a detour. The question is:* which detour? *In theory, the number of detours (think of the possible paths in Pat's search space) is limitless. Without guidance, detouring leads everywhere. Too much guidance leads to more of the same. It's in this area where I see boundaries and constraints as instrumental. If we cannot follow the intuitive, direct, solution path, we need to take the detour—and constraints point us in the right direction, to the right detour.*

Why is less more, but not for everyone? *Let me answer with an example. There's a saying in the Finnish army: if three soldiers cannot move the cannon, take one away. In another army, another soldier is summoned. In a third, the cannon is abandoned. The different responses depend on whether constraints (the immovable cannon, the adequacy of resources to move it) are seen as enabling or disabling* (Gibbert, Valikangas, & Hoegl, 2009).

We can sort these views into four categories, each defined by how individuals respond to abundant vs constrained resources. The Finns are among the

"Resourceful" who make a lot of a little. (Constraints are enabling.) Adding another soldier is the response of the "Resource-Driven" for whom high input leads to high output. (With abundance, there is no constraint). Abandoning the cannon could be the response of two different groups (for two different reasons): the "Resource Victims" who act as if that low input must lead to low output. (Constraints are disabling), and the "Resource Burners" who make too little of what they have (The disabling constraint here is the inadequate response, given the adequate material resources).

So, less is only more if you're in the Finnish Army?

At least when it comes to cannons.

References

Barenboim, D. & Said, E. D. (2004). *Parallels and paradoxes.* New York: Vintage Books.

Cilliers, P. (1998). *Complexity and postmodernism.* New York: Routledge.

Dillard, A. (1989). *The writing life.* New York: Harper & Row.

Gibbert, M., Valikangas, L. & Hoegl, M. (2009). Scarce resources inspire creativity. *MIT Sloan Management Review* online publication. http://sloanreview.mit.edu/improvisations/2009/04/08/scarce-resources-inspire-creativity/.

Hoegl, M., Gibbert, M. & Mazursky, D. (2008). Financial constraints in innovation projects: When is less more? *Research Policy, 37*(8), 1382–1391.

Stokes, P. D. (2001). Variability, constraints, and creativity: Shedding light on Claude Monet. *American Psychologist, 36,* 355–359.

Stokes, P. D. (2006). *Creativity from constraints: The psychology of breakthrough.* New York: Springer.

Stokes, P. D. (2011). Claude Monet 1840–1926. In M. A. Runco & S. Pritzker (Eds.), *Encyclopedia of Creativity* (2nd ed., Vol. 2, pp. 136–139). London: Elsevier.

Contents

About the Authors

This book is a kind of conversation between two collaborators. One is a professor of psychology at Columbia University, New York; the other, a professor of marketing at Lugano University, Switzerland. Each is the primary speaker/author in sections based on their specific expertise; in each section, one comments/expands on/asks questions about what the other writes. As pointed out in the Preface, each "speaks" in a different typeface. Since there is a lot of cross-talk here, our names also appear.

What Do They Have to Say ... About Themselves?

Pat: I'm the psychologist. My expertise is in problem solving and creativity. Most of it was acquired (in art school and advertising) before I became either a psychologist or an academic. Everything I do now comes out of what I did then.

Michael: *What did you do then?*

Pat: I painted and printed at Pratt, wrote advertising copy for J. Walter Thompson in New York and Tokyo, became a creative group head at Ted Bates & Co. I worked with incredibly smart, talented people. We had a lot of fun. We sold a lot of package goods (things that come in packages and sit on shelves).

Michael: *If you were having fun, why did you leave?*

Pat: It stopped being fun. At some point I became (like most experts) stuck in a successful solution.

Michael: *Is that another way of saying "burned out'?*

Pat: No, it means bored. There was nothing left to be learned. Experts get paid to do what they do well over and over again. What I'd learned to do really well (and I had very good teachers) was coming up with what Ted Bates was called a USP, a

Unique Selling Proposition. If well executed, the USP became a claim/a position/a promise that a product (even a parity product) could "own."

Michael: *Can you give me an example?*

Pat: Sure, but let me add something first. This is important: an effective USP didn't come out of nowhere. There was—just as in academia—a great deal of research behind it. Product research (I always talked to the R & D people), market research (competitive promises), and consumer research (focus groups).

An example? Sun-tan lotion. The target audience was teen-agers, so focus groups with teen-agers. Why did they use sun-tan products? To look good, to start a conversation (That's a great tan. What are you using?), or to start more than a conversation (Could you put some on my back?). Nothing useful there. Answers to another question (How could someone get a perfect tan?) were useful. To get a good "base" you had to burn first. You did that with baby oil. Next stop, R&D. Could they put a drop of baby oil into the bottle? Yes? Thank you, I have a USP: Baby oil tan without baby oil burn.

Notice, the USP itself was constrained by both audience (what they believe and what they want) and product (what it can deliver). In turn, the USP became a very productive constraint: its promise was specific and demonstrable. Executions were easy. Like all good ideas, it had "legs," art directors and writers could "run off" with it.

Michael: *Do you think that's true of all good ideas? That others run off with it.*

Pat: Yes, that's how you know.

Michael: *Two more questions: why psychology and why the university?*

Pat: I always read everything I found in the popular press about creativity, and I found it all theoretical, impractical, academic (in the worst sense of that word). I wanted to write something real, practical, and not academic, about creativity. I also always wanted to be a doctor (so the university), and while I knew a lot about psychology, there was a whole lot left to be learned (so psychology).

Michael: *Did you bring anything besides your ideas on creativity to the university?*

Pat: My problem-solving approach. Whatever I work on must, like the USP, be understandable, useful, and generative (meaning it must lead to other useful things). This is of course why the problem-solving literature has been most influential in my academic work.

Michael: *Anyone or anything specific here?*

Pat: Walter Reitman's chapter on the role of constraints in the composition of a fugue. Reitman introduced the idea of paired constraints. In the chapter he describes their roles as limiting and directing search (Reitman, 1965). The only follow-up I found on the dual roles was the other important-to-me paper, Herb Simon's, on the structure of ill-structured problems (Simon, 1973).

Michael: *Last question for now. What is the most useful thing you've done?*

Pat: Several years ago, I would have said developing the paired-constraint model to analyze past, and facilitate future, creations/innovations in the arts and in business. Now, I would say using the model to design a radically simplified math curriculum called *Only the NUMBERS count* (Stokes, 2014a, c, 2016). Yes, I came up with the name (advertising) and, yes, I designed the teaching materials (art). The curriculum uses (1) an English-language version of the Japanese count I learned in Tokyo to solve the place-value problem; (2) a count-and-combine chart to teach single and double digit addition and subtraction in kindergarten and first grade: and (3) a multi-operation chart to teach multiplication, division, and fractions *simultaneously* in second and third. The curriculum meets all my criteria: it is accessible, applicable, useful, and (gratifyingly) successful. I'll talk about it a lot more later, but I think it's time we talked about you, Michael.

Tell us your story.

Michael: *I'm the marketing expert. My research interests are innovation management under constraints, research methods and the roles of similarity and categorization in management thinking. For example, similarity and categorization morphed into themes and taxonomies in my most recent book, Using Thematic Thinking to Achieve Business Success, Growth, and Innovation: Finding Opportunities Where Others Don't Look* (Froehlich, Gibbert, & Hoegel, 2014).

Pat: Where did your questions (about constraints, innovation, themes) come from?

Michael: *Well, a bit like you I started out with the arts, or shall I say the one art that I understand and can practice, cooking. After I passed my A-level exams, I got bored sitting in front of someone holding forth about some kind of topic (my understanding of studying at the time). So I decided to follow in my father's footsteps, i.e. apprentice myself to a master chef. Note, this was back in the early nineties, when being a "chef" was far from being a cool TV celebrity. You worked outrageous hours, were yelled at constantly, and had to do exactly as your boss told you. Put the asparagus on the plate in a slightly different angle? No way!*

But I admired my boss (he had two Michelin stars). I admired his creativity, his being able to work with whatever ingredients were at hand, turning them into something new, something unexpected, every time. As you said, it helps having talented people to work with and work under.

Pat: Tell me about a day in his kitchen.

Michael: *How about every day? Every day, my boss would grab a cup of coffee and disappear (every morning) in the cold room, closing the door behind him. No-one was allowed to disturb him, and some of us made rude jokes about the goings on in the cold room. Every time he emerged (leaving behind his empty coffee cup), looking ever so pleased with himself, and started writing the day's menu (which indeed changed every day).*

Pat: Do you suppose that taking stock of what was left in the cold room (certainly a constraint) was his way of developing new ideas for the menu, creatively using what was there rather than starting from scratch?

Michael: *Exactly, and think about it. Imagine you learned in the early afternoon, while at Barnard College that you had guests coming over for a dinner party—what would make your cooking more creative? Leaving the college immediately, going to the supermarket and buying groceries to cook up something fancy, or (not having the time for shopping) making do with what's at home?*

Pat: As long as there are some things (anchovies, garlic, capers, those kinds of things) at home, making do.

Michael: *Well, there you go. There is something magical about being able to use constraints creatively. If we cannot get what we think we need, we can break through by using the constraints that seemed to hold us back.*

Pat: So, your interest in constraints seems to have started before you became a professor.

Michael: *I didn't know it then, but yes, that's probably true.*

Pat: How did you make the move from resource constrained 2-Michelin star cuisine to the science of innovation.

Michael: *A quote by Paul Cilliers* (1998) *got me started: "boundaries constitute that which is bounded." Constraints in this sense are enabling as well as containing. They have, to parallel your ideas, a dual nature.*

Pat: I want to know more about enabling.

Michael: *You will, later.*

What Do They Have to Say … About Writing the Book?

Pat: The book actually began with Michael (and his co-editors) asking me to write a paper for a special issue on constraints in the *Journal of Product Innovation Management* (Stokes, 2014b). I'd been in business, but I wasn't familiar with the management/innovation literature. That was important. I needed to relate the things

I wrote about to things an audience new to me already knew. I needed a lot of help to cross disciplines. Michael helped a lot.

Michael: *It was so much fun. I had read about Pat's work and realized that we seemed to have worked largely in parallel on the very same topic, and with slightly different points of departure. So I approached Pat, inviting her to contribute a paper. Happily I made it clear at the beginning that the paper would still have to go through the usual double-blind review process (the paper is read and commented on by a number of knowledgeable experts in the field and neither they know who the author is nor does the author know who they are—in theory). In the end Pat sent a paper which looked and read so different from anything I had previously seen in business management. I loved it and eventually, we managed to convince the reviewers to resist their immediate gut reaction of shoehorning it into the usual format or rejecting it.*

Pat, shouldn't we say something about how we're dividing up the writing?

Pat: Sure, I write about the model per se. I apply it to art and to innovative-by-design businesses, successes that were once start-ups. Think Netflix, and Starbucks. I ask Michael lots of questions. (I also edit a lot of Michael's copy).

Michael: *I tie Pat's model to examples from the innovation literature. I also had an idea for expanding the model. The idea is about transitions from existing (i.e., non-innovative) solution paths to innovative ones. It took a lot of talking to convince Pat that the expansion was useful.*

Pat: Useful is the key word. My focus is always on application, not complication. Michael is more theoretically oriented, but I had to agree, the transition idea does expand the model in useful ways. Nonetheless Michael, we do have to explain the model before we can expand it.

Michael: *Agreed.*

References

Cilliers, P. (1998). *Complexity and postmodernism.* New York: Routledge.

Froehlich, J., Gibbert, M. & Hogel, M. (2014). *Thematic thinking.* Upper Saddle River, New Jersey: Financial Times Press.

Reitman, W. (1965). *Cognition and thought.* New York: Wiley.

Simon, H. (1973). The structure of ill-structured problems. *Artificial Intelligence, 4*, 181–201.

Stokes, P. D. (2014a). How early is early enough? Solving the place-value problem in first grade. *The New Jersey Mathematics Teacher, 72*, 30–40.

Stokes, P. D. (2014b). Crossing disciplines: A constraint-based model of the creative/innovative process. *The Journal of Product Innovation Management, 31,* 247–258.
Stokes, P. D. (2014c). Using a creativity model to solve the place-value problem in kindergarten. *International Journal of Creativity and Problem Solving, 24*, 101–122.
Stokes, P. D. (2016). Thinking in patterns to solve multiplication, division, and fraction problems in second grade. *Journal of Mathematics Education at Teachers College, 7*, 1–10.

Chapter 1
Innovation Is a Problem-Solving Process. Part I: Structure

> *Innovations sometimes occur when people are confronted with a problem and scramble to solve it. At other times, they happen when people embrace a visionary goal.*
> Isaacson
>
> *Necessity and opportunity are both problems.*
> Pat
>
> *Boundaries are both problems and opportunities.*
> Michael

Innovation isn't easy. If it were, it wouldn't be a problem (and you wouldn't be reading this book). Since it is a problem, it pays to think about it in problem-solving terms. Pays because thinking this way both simplifies the problem and suggests a way to solve it. Simplification begins with structure, solution with strategy. Chapter 2 introduces strategy. I (Pat) begin here at the beginning, with Newell and Simon's structural innovation, the problem space. (I also begin in the kitchen, "making do" with what's there.)

A reminder: my inputs appear in this typeface, Michael's in *this one*.

Problem Spaces

A problem space is generally defined as a representation of a given problem by its solver (Newell & Simon, 1972; Robertson, 2001). More specifically, it is a *subjective* representation of an *objective* (i.e., given) problem. Subjective means that representations of a given problem will differ depending on whether the solver is an expert or a novice. (I return to this later in the chapter: see Problem Solvers.)

In either case, the parts of a problem space are the same.

- An *initial state*, which represents the current situation.
- A *goal state*, which represents the desired/changed situation. The goal state includes a criterion that defines or specifies the product of the change.

P. D. Stokes and M. Gibbert, *Using Paired Constraints to Solve The Innovation Problem*, https://doi.org/10.1007/978-3-030-25771-2_1

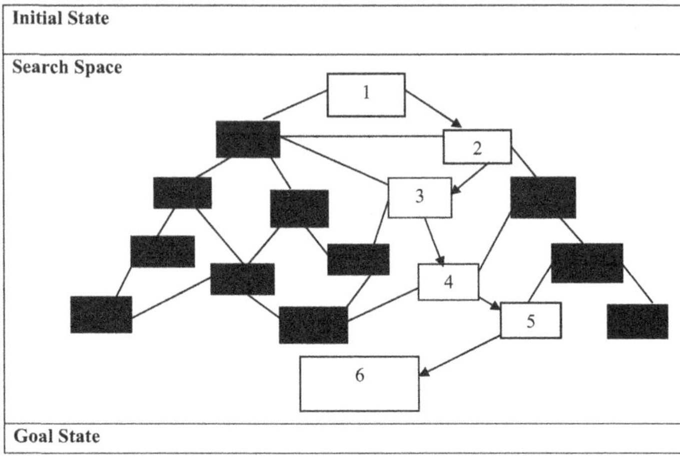

Fig. 1.1 Well-structured problem space. White boxes indicate the solution path

- A *search space* in which a solution path is selected/constructed to transform the initial state into the goal state.

There are two kinds of problem spaces. They differ in how completely specified their parts are. Innovation is only possible in one (not the first one).

Well-structured problems. In a well-structured or well-defined problem, all parts of the problem space are specified. Figure 1.1 shows a well-structured problem. The boxes in the figure represent all possible choices; the lines represent all possible paths. The goal state is a traditional pesto. The initial state represents the uncombined ingredients. The numbered white boxes show my cookbook's six-step solution path for making pesto.

There are six steps in the solution path.

Step 1: Add 2 cups of basil leaves to a food processor.
Step 2: Add 3 cloves of garlic.
Step 3: Add 1/3 cup of pine nuts.
Step 4: Add 1/3 cup of grated parmesan cheese.
Step 5: Add 1/3 cup of olive oil.
Step 6: Cover and pulse until ingredients are combined.

Since all parts of the problem space are specified, there is no need for search. The same is true of all favorite or famous recipes—you don't fool with them. Ditto for assembly lines. Whether their products are cars, computers, or clothing, parts are reliably, repetitively put together in an invariant order. In both examples (recipe, assembly), innovation is not possible. This is true of all well-structured problems. Innovation or novelty is only possible with ill-structured problems.

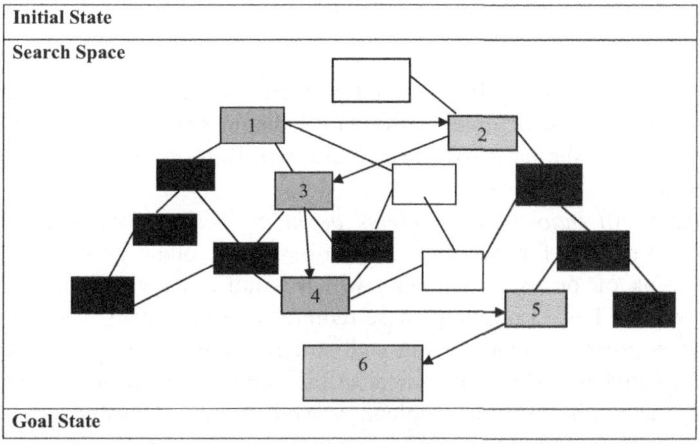

Fig. 1.2 Restructured problem space. Gray boxes represent the new solution path; white boxes represent precluded steps from the previous solution path

Ill-Structured Problems. If a problem is ill-structured, not all its parts are specified (Simon, 1973). To keep things simple, let us suppose that I am missing three ingredients required in my pesto recipe, which is now the initial state. (Think of this as a resource scarcity problem.) Scarcity makes my problem space ill-structured. Search, both literal and figurative, and re-structuring are necessary. To solve the problem, I search in my refrigerator. (Think of this as bricolage.) I am in luck, my search yields parsley, walnuts, and Pecorino Romano cheese.

Figure 1.2 shows my restructured problem space. The white boxes show the parts of the previous path that were replaced; the gray boxes show the new solution path. It includes substitutions at steps 1 (parsley for basil), 3 (walnuts for pine nuts), and 4 (Pecorino Romano for parmesan). The complete solution path, like that in Fig. 1.1, has six steps

Step 1: Add 2 cups of *parsley* leaves to a food processor.
Step 2: Add 3 cloves of garlic.
Step 3: Add 1/3 cup of *walnuts*.
Step 4: Add 1/3 cup of grated *Pecorino Romano* cheese.
Step 5: Add 1/3 cup of olive oil.
Step 6: Cover and pulse until ingredients are combined.

The pesto problem provides us with a preview of two important things. First, it is a much simplified example of *solution-by-substitution*. Solution-by-substitution is the strategic core of the paired constraint model (detailed in Chap. 2). Second, it demonstrates the *paradoxes* present in all innovation problems.

The Paradoxes in Innovation Problems

Novelty is only possible in ill-structured problems. It follows then that innovation problems must be ill-structured. However, all begin as well-structured problems. This is only one of the paradoxes. There are two. Both involve re-structuring of a problem space.

Paradox 1*. All innovation problems begin as well-structured problems.* All. There are two reasons for this. One is brain-based; the other, problem-based. First, you can't think of, or about, anything which is not already in your brain's associative networks. Two, a problem space requires an initial state. The initial state in an innovation problem is an existing well-structured problem, the choice of which (as a later chapter will show) is critical to the innovativeness of the solution.[1]

To become an innovation problem, a well-structured problem must require (necessity) or suggest (opportunity) changing/excluding/precluding something or somethings. Excluding parts of a well-structured problem makes it *ill-structured.* Re-structuring not only permits, but requires, novelty.

Paradox 2*. Solving an innovation problem makes it well-structured.* Now novelty is no longer possible, unless (1) you begin again at the beginning and restructure your problem space to make it ill-structured, or (2) your problem is only *partially solved.* Partially solved means that your solution path is permeable, or your goal criterion is generative. A permeable path permits multiple substitutions.[2] A generative goal criterion begets successive sub-goals. In other words, your solution permits or suggests its own re-structuring.

Given the paradoxes, we thought about some of the possibilities.

Innovating as Re-Structuring: Four Possibilities

When is innovation possible? The obvious answer is when a well-structured problem is made ill-structured. In this section, we talk about four general ways in which this can happen. In Chap. 3, we present a complete typography.

Possibility 1*. Substitutions are necessary.* Parts of an original solution path are no longer available. Necessity has turned a well-structured problem into an ill-structured (because incompletely specified) one. My pesto problem fits in this

[1]It is also what Michael calls a boundary: it specifies the domain in which the solution process proceeds.

[2]Some solution paths are too permeable. Imagine the shampoo aisle in your drugstore. There are X brands, each with at least Y versions of basically the same produce. Look at the hot cereal section in your supermarket. How many varieties of instant oatmeal are there? Multiple choice has become imitative, not innovative.

category. The goal criterion is the same (sauce for pasta/spread for bread). A few substitutions (parsley, walnuts, Pecorino Romano) will complete the new solution path—and make the problem well-structured.

Michael, you're the chef. What would you add to my pesto example?

A dash of analogical reasoning and a cup (a full cup) of prior knowledge. In this case, prior knowledge of the basic ingredients and fundamental mechanics of making pesto is critical. What we need are nuts, grated cheese, and some kind of herb and of course, garlic (no substitute for garlic, unfortunately). Without prior understanding to extend to a new situation, we will not be able to substitute.

Let me give you another example. At the Max Planck Institute in Leipzig Germany, creativity in chimpanzees is studied (following Koehler's classic studies of about 100 years ago). In one memorable experiment, the chimp is given a couple of peanuts in a very long, very thin plastic tube. The poor chimp can't get to these peanuts but you can see that she desperately wants them. "Hold on," she says (or so it seems), "peanuts float on water, right? I've seen it happen many times in my water bowl. Unfortunately, there is no water in the tube, only air, and peanuts do not float on air. Wait! Maybe I can substitute the air in the tube with water!"

This is the solution. But how did she get the water from the bowl into the tiny tube?[3] *By repeatedly sucking up the water and squirting it into the tube with her mouth until she actually got to the peanuts.*

Possibility 2*. Substitutions are desired (not required).* The difference from Possibility I is that here permeability becomes opportunity. There are multiple substitutes for basil (arugula, spinach, baby greens), for pine nuts (almonds, hazelnuts, pecans), and for parmesan (manchego, asiago, pecorino) that can satisfy the goal criterion for pesto (and perhaps turn my kitchen into a small business). The same is true of classical forms in music: multiple solution paths can satisfy the criterion for fugue, sonata, or variation, especially for variation—think of Mozart's variations on *Avez direi-je mamam?* (aka *Twinkle, Twinkle Little Star*). Again, the goal remains the same, only the solution path differs. Again, the solution restructures the problem, making it once again well-structured.

Again, the basic mechanism here is analogy. Except that here analogy is used as a communication device for making new ideas more readily understood by an audience with prior knowledge of the concept on which the analogy is based. To stick with the culinary examples you provide, Pat, innovations in Italian cuisine are usually explained using this substitution technique. For instance, "carpaccio" is anything cut up thinly (it started out as beef carpaccio but has made forays into anything from pineapple to red beets). Similarly, "cappuccino" [the milk-coffee amalgamation we used to consume before the "Grande Latte" became (in)famous]

[3]Those chimp scholars are really mean: the tube has a minute diameter and is so long that it is impossible to simply pour the water into the tube.

is called thus because it looks like a young Cappucchine monk's neatly shaved head from above (white with a rim of dark brown).

Possibility 3*. Opportunity or aspiration initiates a new goal.* Now novelty is required in both path and goal. In this case, the new path simultaneously specifies and realizes the new goal.[4] By specify, I mean define or identify the goal criteria. By realize, I mean accomplish or meet it. I'm thinking about Mozart and Monk here. Mozart's variations did not alter musical form. The criteria for variation remained the same, only the path differed. Monk's variations altered both form and path. Sparse (silences occur and reoccur), sequential paraphrases (successive transformations are themselves transformed) of an initial theme redefined and restructured improvisational form in jazz. I'm also thinking about Starbucks, but I'm saving that for another chapter.

*For another example, consider the idea of selling a song without music, and let's see how the new path simultaneously specifies the new goal. In the 1950s jukeboxes became popular. Columbia pressed a three-minute "no-song" song on their LPs and had people actually pay for it (as in, pay for listening to silence). Now why would a major record label spend resources to see music without music? Problem was (sorry, I meant opportunity) that these jukeboxes became victims of their own success. Or rather, some diners were victims of jukeboxes' success—they just wanted to dine in peace and quiet (or perhaps listen to classical music which was unavailable). So essentially, the new "song" (the new path) realizes the new goal of providing some moments without the jukebox playing Elvis in the background. All of this sounds pretty much like an April Fool's joke, you reckon? In fact, this "no-song" turned out to be one of their most-played songs (*http://www.stumblerz.com/silence-is-it-gold/*). There's a great analogy in the now defunct iTunes, which I'll talk about in Possibility 4.*

Before we get there, how about selling tickets to hear classical music without music? I'm thinking of John Cage's (in-famous) composition *4'33"*. The title is the content: 4 min and 33 s of silence during which the pianist sits (just sits) at the piano. Cage's idea was that any sound (here the ambient sounds in the concert hall) can be heard (re-categorized) as music.[5]

Possibility 4*. Generative goal criteria.* Generative means suggestive of its own development. Go back to the section on Monet in the Preface. Chevreul's color wheel presented Monet with a new problem: show how light breaks up. The problem became his overall goal. It also generated three successive sub-goals: show how light breaks up on things, between things, and finally (on the surface of the pond) by itself. Each successive solution path made Monet's problem well-structured. Each new sub-goal made it ill-structured. This is rare. It represents re-structuring at its highest level.

---

[4]This co-occurrence of path completion/goal designation also appears in Duggan's strategic intuition model. He puts it this way, "the achievement and the goal arise at the same time" (Duggan, 2007, p. 23).

[5]This is akin to Merce Cunningham's idea that any movement can be choreographed as dance (Stokes, 2016).

As promised, the iTunes example. Columbia had already (1950s) sold a song without music. iTunes' version (2011) appears to be the same thing: music that has no music and is best described as silence. Was it? Were the two silences the same? In the Columbia version, silence was an end in itself. Diners just wanted a moment of silence, rather than another round of Jailhouse Rock. This use of silence, while innovative, is not "generative." It is a means to an end, not an end in itself. In the iTunes example, the silence is no longer simply silence but a vehicle for making a donation to Caritas. The goal criterion here (making a donation) prompts a reinterpretation of the song without song as a one-minute reflection about Haiti and the need for help. It took sixty years, a massive earthquake, and the creative energies of Caritas, a major music label, and one of the world's most innovative advertising agencies (Jung von Matt/Limmat) to re-categorize the absence of a song where a song should be and to make consumers pay for it.[6]

Michael, I have a question. Isn't this re-categorization idea related to what you called an "ad hoc" category?

Not me, a cognitive psychologist, Barsalou (1983). *An ad hoc category is based on the function a product performs at the moment it is observed. In the iTunes example, the function of the silence is to draw attention to something other (need/a donation) than sound. A more mundane example involves herbs growing in pots on my balcony. They need water. I have no watering can "at hand," but I do have a teapot. If I use the teapot, it will effectively be re-categorized as a watering can, and not simply as a teapot. Another Barsalou example is a basketball, a product not usually put into the category "lifesaver." However, like lifesavers, basketballs float on water and can be quite useful after a ship-wreck.*

I like the idea of re-categorization. It offers solutions to problems in the "substitutions are necessary" category, like watering plants or surviving a ship-wreck (which moves teapots and basketballs to Possibility I).

How Innovative Is a Re-Structuring?

To answer, we borrow criteria from the creativity literature (Stokes, 2014). There are three, all require novelty plus something else. Using these criteria, a solution may be:

Novel and useful. This is the most common level: the innovation solves the problem at hand. How innovative here translates (nicely for measurement purposes) into how useful? How often used? How many users? My pesto problem and Michael's three minutes of silence (a break before the next song) belong here.

Novel and generative. Generative means that the innovation propels its own development by suggesting subsequent problems to solve. For example, goal criteria (like Monet's "show how light breaks up") that generate successive sub-goals

[6]The revenues for iTunes one minute of silence are donated to help the people of Haiti.

(on things, between things, by itself) or that use one innovation (Columbia's silence instead of song) to create another (iTunes reflection/donation instead of song).

Novel and influential. This is the least common, most innovative level. Here the innovation expands or, in the rarest cases, creates a domain.[7] Influence means that the solution changes the way that others look at, think about, and—most importantly—solve problems in the domain. Because Monet's innovation (Impressionism) expanded its domain, influencing the way other artists solved (and still solve) their painting problems, it also belongs here.

Pat, when I think about Impressionism, I also think about resource constraints. For example, time constraints required painting quickly and in bold strokes to capture the fleeting moment. A good business example, where resource constraints created a new category, are tablet computers. The constraint here was weight rather than time. Tablets, previously called "netbooks," are scaled down portable computers. Originally designed as cheap play-and-learn computers for kids, they were quickly appropriated by adults and repositioned as very portable, very lightweight computers. The ASUS EeePC701, released in November 2007, redefined an entire industry.

Summing Up: Structure

Novelty is only possible when a problem is ill-structured or incompletely specified. Not surprisingly, innovation problems must be ill-structured. Surprisingly, all begin and—when solved—all (but a few) end as completely well-structured problems. Two exceptions (permeable paths, generative goal criteria) have been noted. The problem space itself is important as a site/template for structuring and re-structuring innovation problems. The next chapter (after a short digression) focuses on *solution-by-substitution*, our strategy for breaking a problem down into the structure of its solution.

Digression I: The Problem Solvers

Chance favors only the prepared mind.
Pasteur

The prepared mind notices what is needed.
Pat

Prior knowledge is preparation for new.
Michael

[7]A domain is defined as an established area of expertise. It is the boundary condition against and within which the innovator works.

The Subjectivity of Problem Solvers

At the start of this chapter, problem spaces were defined as *subjective* representations of *objective* (i.e., given) problems. Subjective means that representations of a given problem differ depending on whether the solver is an expert or a novice, and on the domain of the solver's expertise.

Subjectivity and Attention. One way in which subjectivity affects a representation involves *attention*. Because they know specific things, experts in one area/domain direct their attention to different aspects of the same situation than either a novice or an expert in another area (Chi, Glaser, & Farr, 1986; Ericcson, 1996). For example, imagine three experts and one novice looking at a painting by Matisse. The painting I am thinking of is called *The Goldfish* (1912). The simplification (mine) shown in Fig. 1.3 emphasizes Matisse's: the three outlined, flatly colored, orange shapes in the bowl are signs, icons of the objects (the actual goldfish) that they represent.

One of our experts is an art historian. He is thinking about Matisse's three stylistic phases. Which is represented by *The Goldfish*? (It's the middle period when the iconic was introduced). Our second expert is a dealer. She is thinking about/calculating the value of the painting. How much is a painting like this one worth? (A lot). Our third expert is a painter. She's thinking about what she can borrow from Matisse to solve her painting problem. (Yes, Matisse is influential. Yes, she'll borrow the limited palette). Our novice, of course, knows nothing about art history, marketing, or making. He is simply thinking about whether or not he likes the painting. (He likes it enough to look a little longer.)

Fig. 1.3 Schematic drawing of *The Goldfish*, 1912

Subjectivity and Chance. Let me linger a bit with our painter. She has noticed something (the limited palette) that she needs. This is something I've noticed repeatedly—once you start working (on an advertising campaign, a painting, a paper), things that you need (phrases, images, connections) somehow come into sight, into mind. This does not happen with novices, who do not know what they need. Because they do know, experts do notice.

In his book on *Creative Strategy,* Duggan (2013) discusses a variant: noticing "precedents," things that have helped solve similar problems. For example, one precedent critical to the innovation known as Netflix was the flat membership fee charged by Reed Hastings'[8] gym. The flat fee was noticed after Hastings had to pay a large late-return fee to Blockbuster. With a small change (the flat fee is charged monthly, not annually), Netflix precluded Blockbuster's rental-plus-late-fee payment plan and promoted its newly noticed and modified flat-fee version.

Pasteur was right, chance does favor the prepared mind (Austin, 2003).

Michael was right: no prior knowledge (the preparation), no noticing (and no new knowledge).

Subjectivity and Efficacy. Noticing different things is just one effect of expertise. The other—knowing more things, many more things, having a very big tool box[9]—increases the number of possible combinations, substitutions, structurings, and re-structurings. In short, experts can construct more elaborate and more efficient problem spaces than novices (Weisberg, 2006). When experts collaborate with experts (with different kinds of expertise), tool boxes (all that borrowing) and possibilities (all those combinations) get even bigger.[10]

Caveat: not all experts do this. Tom Kelley of IDEO, who prefers "tinkerers" to experts gives us one reason: "Expertise is great until it begins to block you off from new learning" (Kelley, 2001, p. 181). I can offer another: many, if not most, experts get *stuck in successful solutions*. Experts are rewarded, well and reiteratively, for solving specific problems (think of an intellectual property attorney: if your property isn't intellectual, you hire a different lawyer) or for producing recognizable solutions (think of a collector: if you buy a Matisse, you want it to be recognizably a Matisse).

This is why innovation depends on constraints—they preclude currently "successful" solutions (even, and especially, if the solution is the innovator's own).[11]

---

[8]Hastings is the founder of Netflix.

[9]See *Digression II: Thinking Inside the (tool) Box.*

[10]Isaacson offers multiple examples of effective/critical collaboration in his book *The Innovators…* (2014).

[11]As usual, I am thinking of painters (Monet, Matisse, Chuck Close for starters) who restructured their own innovative solution paths more than once (Stokes, 2014; Stokes & Fisher, 2005).

References

Austin, J. H. (2003). *Chase, chance, and creativity: The lucky art of novelty.* Cambridge, MA: The MIT Press.

Barsalou, L. W. (1983). Ad hoc categories. *Memory & Cognition, 11,* 211–227.

Chi, M. T. H., Glaser, R., & Farr, M. J. (1986). *The nature of expertise.* Hillsdale, NJ: Erlbaum.

Duggan, W. (2007). *Strategic intuition: The creative spark in human achievement.* New York: Columbia Business School Publishing.

Duggan, W. (2013). *Creative strategy: A guide for innovation.* New York: Columbia Business School Publishing.

Ericcson, K. A. (1996). The acquisition of expert performance: An introduction to some of the issues. In K. A. Ericcson (Ed.), *The road to excellence: The acquisition of expert performance in the arts and sciences, sports and games* (pp. 1–50). Mahwah, NJ: Erlbaum.

Isaacson, W. (2014). *The innovators: How a group of hackers, geeks, and geniuses created the digital revolution.* New York: Simon & Schuster.

Kelley, T. (2001). *The art of innovation: Lessons in creativity from IDEO, America's leading design firm.* New York: Doubleday.

Newell, A., & Simon, H. (1972). *Human problem solving*. Englewood Cliffs, NJ: Prentice-Hall.

Robertson, S. I. (2001). *Problem solving*. New York: Psychology Press.

Simon, H. (1973). The structure of ill-structured problems. *Artificial Intelligence, 4,* 181–201.

Stokes, P. D. (2014). Thinking inside the box: Creativity, constraints, and the colossal portraits of Chuck Close. *The Journal of Creative Behavior, 48,* 276–289.

Stokes, P. D. (2016). *Creativity from constraints in the performing arts*. New York: Stokes.

Stokes, P. D., & Fisher, D. (2005). Selection, constraints, and creativity case studies: Max Beckmann and Philip Guston. *Creativity Research Journal, 17,* 283–391.

Weisberg, R. W. (2006). Models of expertise in creative thinking: Evidence from case studies. In K. A. Ericcson, N. Charness, P. J. Feltovich, & R. R. Hoffamn (Eds.), *The Cambridge handbook of expertise and expert performance* (pp. 761–788). New York: Cambridge University Press.

Chapter 2
Innovation Is a Problem-Solving Process. Part II: Strategy

It is necessary to create constraints in order to invent freely.
Umberto Ecco

It is necessary that the constraints be paired in order to innovate freely.
Pat

It is necessary to see boundaries as paired constraints, both precluding and promoting.
Michael

For the writer of fiction, Ecco (1994) continues, "the surrounding world [the one created by the writer] provides the constraint." His characters are obliged to live according to its laws, making the narrator himself "the prisoner of his own premises."

For the writers of this book, Ecco's "premises"—his surrounding world—are not solitary, but rather …

- In Michael's wording, a boundary both constrains and enables. *There is something magical about constraints in that they take us out of our comfort zone, which typically operates in spaces where we can procure what is needed to solve a problem (at least in developed countries). In constrained spaces, the opposite is true.*
- In Pat's wording, the product of paired constraints that preclude other worlds (other times, other languages, other customs, other rules) and promote this world (this time, etc.).
- In other words, by precluding other worlds, the boundary enables the author to invent freely in *this* world.[1]

[1]The premises in *The Name of the Rose* are bounded by time (November, 1397) and place (a monastery in Avignon). Multiple constraints determined both. Its protagonist, for example, was to be "an investigator, English if possible… with a great gift of observation and a special sensitivity in interpreting evidence." "These qualities," Ecco tells us, "could only be found among the Franciscans, and only after Roger Bacon" (Ecco, 1994, p. 26).

P. D. Stokes and M. Gibbert, *Using Paired Constraints to Solve The Innovation Problem*, https://doi.org/10.1007/978-3-030-25771-2_2

For inventors of all kinds, paired constraints are tools that break an innovation problem (here, Ecco's novel, *The Name of the Rose*) down into the structure of its solution. The pairs work this way. One identifies a specific element in an existing well-structured problem to be precluded. The other identifies/promotes a substitute. The process is self-sustaining: one substitution suggesting or requiring another. It is also incremental. Step by step, substitution by substitution, the constraint pairs construct a new solution path. This is why the process is called solution-by-substitution.

Importantly, the new solution path *is* the innovation.

The Strategy: Paired Constraints

In economic terms, constraints are one-sided, connoting limitation, restriction, or scarcity.[2] In contrast, the problem-solving literature considers constraints two-sided, *pairs with different purposes*. One of each pair satisfies the dictionary definition: it precludes something specific in the current situation. The other expands the definition: it directs and promotes search for a substitute. Whether by necessity (scarcity) or by choice (opportunity), the precluding starts the process.

There are four constraint pairings: *source constraints*, which define domains; *criterion constraints* which stipulate goals; *task constraints,* which govern materials and their applications; and *subject or category constraints*, which specify the kind of innovation. We elaborate on each in turn.

Pair 1: Source Constraints. Source constraints are specific to domains, defined as specialized areas of expertise with agreed-upon performance and stylistic criteria. They are the "source" of criterion, task, and subject constraints. Once mastered, they provide elements for an innovator to work against (preclude) or with (promote). Larry Rivers, a painter and jazz musician, called them the "first chorus" on which the performer or innovator improvises (Rivers, 1987). Improvisation, like all innovation, depends on mastery. Novices not only have their constraints chosen by teachers or mentors, they may not know if something is new or not. Mastery allows experts, who do know, to select their own constraints.[3]

Pair 2: Criterion Constraints. A criterion is a recognized, agreed-upon standard that identifies a product, process, or style. The initial state in a problem space represents the preclude half of this pairing. It is the currently accepted standard that an innovator aims to resolve (in the case of necessity) or to replace (in the case of opportunity) to replace. Necessity usually implies scarcity. An existing, still desired, criterion can no longer be satisfied. Here, to resolve literally means to *re-solve*. In this case, there is no pairing: the goal is realizing the established criterion via a novel solution path.

---

[2]Whether or not such limitations impede or expedite innovation (Baker & Nelson, 2005; Cunha, Rego, Oliveria, Rosado, & Habib, 1999; Gibbert, Hoegl, & Valikangas, 2007; Katila & Shane, 2005) will depend on their scale and their interactions with other variables, including how innovative teams perceive material resources (Weiss, Hogel, & Gibbert, 2014).

[3]Choosing your own constraints is how Pat defines "artistic freedom" (Stokes, 2006).

In the case of opportunity (or aspiration), there is a constraint pairing. The criterion for the initial state is precluded; the criterion for the goal state, which defines the innovation, is promoted. Here, both criterion and solution path are new. As we will see, the new criterion is only fully specified when the new path is fully specified.[4]

Pair 3: Task Constraints. Task constraints are concerned with materials and accepted methods (skills, procedures) of using them, and asking the question what *else* can we do with them? They are what I call *the basics* in an expert's *tool box* (see Digression II). They come in two kinds that together constitute expertise in a domain. The first involves *semantic* memory, what you know. Importantly, what you know has a lot to do with what you notice. The second involves *procedural* memory, what you can do with what you know.[5]

Pat, there's something I want to add about task constraints and innovations. I'm thinking of task constraints as points of departure: the destination being different from the destination suggested by the departure. I'm thinking of equifinality, achieving different ends with the same means, and also of virtuosity, using tools smartly (in masterful, yet surprising ways).

Michael, I like *points of departure* because they point back to noticing: noticing what's needed or what's happening at a specific point in solving a problem, particularly at the point when one type of problem *transitions* into another, the first problem providing the core or catalyst for the second, more innovative destination.

One more thing, analogy, achieving different ends with similar (but not identical) means.

Pair 4: Subject or Category Constraints. These are the simplest to describe. They are concerned with classifications. What style of painting: Fauvist, Cubist, Minimalist? Which kind of novel: historic, romantic, fantastic? What category of car: compact, sedan, SUV? Powered by: gas, electric, both? What sort of car service: chauffeur, taxi, jitney, Lyft, self-driving car? What type of watch: analog, digital, battery, winding, self-winding? (Yes, winding and self-winding. Watches are one of several high-end categories where old is new again.).

Truly new categories are rare. Uber is a gig-economy variant of car and driver for hire. Lyft is a variant of Uber. Airbnb is a gig-economy variant on traditional b&bs (bed-and-breakfast). *Couch surfing (i.e., offering one's couch, sofa-bed, or straight sofa to paying guests[6]) is the cheapest version of b&b—or is it? According to my consumer behavior students, it is a convenient way of finding a significant other, the ideal excuse for a holiday flirt.*

More "truly new" are categories related to eating. Companies like Blue Apron, Home Chef, and Plated deliver pre-measured (exactly a pinch of saffron) ingredients and directions for combining and cooking. (Are these a variant on cooking

---

[4]Ahuja and Katila (2004) describe similar paths, albeit with different terminology. In their usage, path-deepening search resolves an old problem in a new way (this is akin to realizing an existing criterion); path-creating search solves a new problem (this is akin to specifying a new criterion).

[5]This combination of knowledge and skill has not gone un-noticed in the innovation literature. For example, see Miner, Bassoff, and Moorman (2001) on stored knowledge and skills.

[6]See www.couchsurfing.com.

classes, or a way to avoid learning how to cook?) Companies like foodora in the EU or Grubhub in the USA deliver meals prepared in "ghost kitchens." Michael says these businesses are changing the entire value chain.

The Process: Solution-by-Substitution

Pat's Example: Peasant Pesto

To simplify the process, I'm going to elaborate on my pesto problem (Chap. 1), showing how the substitutions were selected.

Step One: Specifying the Initial State. The initial state of an innovation problem is the product, style, or process to be resolved (in the case of necessity) or replaced (in the case of opportunity). All the parts, including the ordering (assemble before pulse), should be specified. Identifying *everything* in the initial state is important for three reasons:

- First, it *provides a template* of where you started/what you are working against. If yours is a problem of necessity, the goal will remain the same. Only the solution path will differ. If yours is a problem of opportunity, both goal and path will differ.
- Second, it *prevents overlooking* some taken-for-granted elements in the initial state, the order in which the elements are assembled. Both elements and ordering are part of the solution path. Some will be retained. These will be part of your new solution path.
- Third, and critically, it *supplies the preclude column* of your innovation problem. Without a preclude column, you have no place to start the substitution process.

Step Two: Re-structuring the Initial State. This is where the paired constraints come into play. To start re-structuring, divide the search space into three columns. Under the one-headed preclude, put the solution path from Step One. What happens next depends on whether your problem is one of necessity or opportunity.

If your problem is one of scarcity, the next step is simple. Identify an element that needs to be re-sourced/replaced. Move it to the top of the preclude column. Notice that your problem is immediately ill-structured. Innovation (to some degree) is now possible. Select its substitute and put at the top of the promote column. Does the substitution suggest another? Iterate until you have a new solution path to the initial goal.

If your problem is one of opportunity, the next step is more complicated. You need to re-specify the goal. Your first pass will probably be too general (a new X, a better Y, a more efficient Z). You will surely construct several possible paths.

Table 2.1 Ill-structured innovation problem

Initial state: Classic basil pesto			
Search space	**Constraint pairs**		
Retain	**Preclude**	→	**Promote**
Garlic	Basil		Parsley
Olive oil	Pine nuts		Walnuts
Pulse in processor	Parmesan cheese		Pecorino Romano cheese
Goal state: Peasant pesto			

Some will be precluded. That one that prevails will specify the goal. The good news is that you will know when it happens. *Defining and accomplishing a goal occur simultaneously.*

Table 2.1 shows the re-structuring of my pesto problem. This search space is different from others in the book. It is the only one in the book with three columns.[7]

- The one-headed ***Retain*** includes those elements of the initial problem space that were not (indeed could not be) precluded. Pesto requires garlic and olive oil and pulsing to turn the ingredients into a paste.
- The one-headed ***Preclude*** indicates the elements which could be replaced.
- The one-headed ***Promote*** shows the substitutions. The promote column plus the retain column is the complete solution path that characterizes and creates (it is the new recipe) my innovation, peasant pesto.

Where did peasant pesto come from? I'm pretending that this was actually not a scarcity problem. I had basil, but I was bored with *classic* basil pesto. I began by precluding basil—this made my problem ill-structured—and continued by substituting parsley for basil. Parsley is less aristocratic than basil, so I began looking for other more *peasant* (and also less expensive) substitutions. My constraint pairs cascaded—step by step, walnuts were substituted for pine nuts, pecorino for parmesan.[8] And voila! Peasant pesto!

Michael's Example: Feature Deletion and Re-categorization

A classic example for the solution-by-substitution is portable music, and the device that started it, the Walkman. Those readers who still remember the late 70s will remember Sony's "Walkman" as the first device that enabled us to privately listen to our favorite tunes wherever we are (for those who do not: the Walkman was a scaled-down version of a cassette player, delivering the sound via earphones). I am

[7]To emphasize the substitutions that create an innovation, subsequent problem spaces will only include preclude and promote columns.

[8]We could also add kinds of constraints to the pesto problem. The initial and goal states represent the criterion pair; the remaining are all task constraint pairs (concerned with materials, here ingredients, and how they are manipulated, here by pulsing).

saying "privately" as portable transistor radios of course were invented (again by Sony, by the way) decades before the Walkman, and they were small enough to be carried around (remember these "Ghettoblasters"?), inflicting your music on everybody within 100 yards.

How did the Walkman come about? Sony produced something very similar, the so-called "Pressman," or a small portable tape-recording device for journalists and others who had to record interviews and such. The idea was quite simple: get rid of that red recording button. Now what IS a recording device that cannot record? There is still the play-back option of course, why would one use ONLY the playback option? To listen to what is recorded on the tape, of course. Now, what is recorded on the tape? To facilitate music consumption, another feature from the Pressman had to be changed: the size of the tapes (the Pressman used tiny specialized tapes which would not fit the conventional Hi-Fi phono-cum-cassette-cum-radio contraptions which used to adorn our living rooms). Thus, the Walkman used conventional tapes consumers were used to. But wait, where does the substitution come in? To actually be able to listen to the music tape (the apparatus is too small to house large loudspeakers and the batteries needed to produce music for public consumption), one needs an additional piece of equipment: earphones!

In short, a solution path using paired constraints (no recording button, device too small to produce loud music, tapes too small to fit the equipment consumers already owned) with a solution-by-substitution (headphones substitute loudspeakers) led to an entire lifestyle we cannot imagine ourselves to be without.

This example is too good not to get a problem space, so I have made one.

Thanks (Table 2.2).

Closing Thought: The Importance of the Initial State

Michael's example is perfect, pithy, and—like my pesto example—emphasizes the importance of the initial state. To reiterate, the initial state (the Pressman) provided the critical preclude column. Without a preclude column, there is no place to start

Table 2.2 Pressman to Walkman

Initial state: Pressman—portable recording tool			
Search space	**Constraint pairs**		
	Preclude	→	**Promote**
	Record and play back		Only play back
	Specialized tapes		Standard tapes
	Speakers		Earphones
	Public listening		Private listening
Goal state: Walkman—portable, private listening tool			

the solution-by-substitution process. Often you will have no choice in your initial state. If you do, choose carefully. The initial state will determine how innovative your solution-by-substitution path will be.

Digression II: Thinking Inside the (Tool) Box

> *Every intelligent painter carries the whole culture of modern painting inside his head. It is his real subject, of which anything he paints is both a homage and a critique...*
> Robert Motherwell (1992)

> *The tool box I'm thinking about is the one inside your head, the one you can't think outside of.*
> Pat

> *Paradoxically, scarcity can make your tool box bigger than abundance can.*
> Michael

When Motherwell (an Abstract Expressionist) says homage, I think retain and when he writes critique, I think preclude. What he's saying is that your domain, your area of expertise, provides both what you work with (the homage) and what you work against (the critique). Both are what I call the basics in your tool box, the one inside your head, the one you can't think outside of. There are also borrowings, things from other domains, things noticed when needed.

As previously, Pat's thoughts are in this typeface; *Michael's are in this.*

The Contents of the Tool Box

The Basics. The basics are the first things that go in. They define your expertise, which means they are domain-specific. They come in two kinds: tools in your head and tools in your hands. The tools in your head involve *semantic* memory (what you know); the tools in your hands represent *procedural* memory or skill (what you can do with what you know). What could we expect to find in Motherwell's tool box? Things that painters know (about styles and materials) and techniques painters use (to produce variations of those styles, with those materials). Michael, help me here, what would we find in a manager's tool box?

Let's use one definition of management: resource allocation. That is, moving money (that's easy to move) or raw material and equipment (that's more difficult to move) from where it is stored to where it is needed. What would we expect to find in a manager's toolbox, then? Probably simple heuristics for deciding just how many resources to move so that teams (that's mostly who does the work) can do the job they are supposed to be doing. The common denominator among these tools is "more is more," or at least "adequate is good." In other words, when it comes to resource distribution, the idea is that where resources are needed, they need to be made available.

How can an innovator make a basic tool box bigger? There are two ways. One is *deliberate*, expanding your expertise in domain-related ways.

You are a painter, a colorist, not a draftsman. You want to look at the world as black and white and in lines. You decide to study under a master printmaker. You are YoYo Ma. You want to learn new ways of hearing and making music. You set off on the Silk Road. *You are Johannes Gensfleisch Gutenberg, and you grow up in the goldsmith trade in a town on the River Rhein surrounded by vineyards (and, of course, wine presses).*

You are a manager. You want to make sure that the work gets done by allocating resources. Where the world for painters is black and white and in lines, the world for managers is red and black. You are trained at university and perhaps later at business school to think that lack of resources (the red figures) is counterproductive and to be avoided precisely by making resources available where they are needed. The deliberate way to increasing the size of the toolbox is firmly tied to a specific discipline. They are happening from inside that domain, and are a means to an end, with the end being an improvement (often in other disciplines) using principles of that, original, discipline.

This somehow sounds deprecating. It isn't meant to be that way; for domain-specific innovations (in particular those which "define" the modus operandi of a particular domain) can hold the clue to radically innovating other domains. The screw press for extracting juice from grapes and olives has been invented in the first century AD, and remained virtually unchanged until modern times (as an aside, I have "inherited" such a press, probably dating from the nineteenth century from the previous owner of my house in Piemonte, and it still works). By the time Johannes Gutenberg used the screw-press principle to innovate the domain of printing, this particular innovation was already some five-hundred years old. Using the same principle for printing, rather than pressing, and in combination with movable type led to what is widely considered the single most important innovation in the second millennium, laying the foundation for separating knowledge from its physical carrier, a process that led to the Renaissance, Reformation, the Age of Enlightenment, the scientific revolution, and the knowledge based economy we are still busy with today. Thus, if you function in the deliberate mode of innovation, you, well, liberate domain-specific (and often domain-defining) principles from the source domain to innovate the new domain.

The other way is ***automatic***, it happens whenever a problem is solved. Every successful substitution, every newly constructed solution path adds new skills—a new way of making and a new way of noticing connections—to a solver's tool box. Monet's first substitution series (which gave us Impressionism) gave him new ways to paint (using contrasting hues rather than values, separate mosaic-like, instead of continuous, strokes) and new connections (the envelope or atmosphere that surrounds, immerses, and connects everything in its colors) that *over time* (over 20 years) led to his second substitution series (changed conditions rather than motifs, shared instead of local colors, etc.). Michael, you have much more to tell us about "over time," don't you?

Yes. The additions and the realizations that "go together" add up and previously separate functions have an almost magical tendency to merge over time making innovation often (very often, most often) incremental. Consider the telephone. We moved from a wall-mounted technology with a speaker (attached to the wall) and a headphone (attached to a wall-mounted equivalent with a wire) to a table-top device which now united the headphone and the speaker into one hand-held unit, which however was still connected to a kind of receiver which remained on the table and which we used to dial telephone numbers (first via a circular device and later via a keypad). The next step up were wireless home phones, which united, in the speaker/headphone also the keypad, and not too long ago, we still had mobile phones which folded up. Camera-phones where the first step in combining the phone with a completely new (and yet, in hindsight, intimately related) function, namely communicating with pictures, in addition to communicating via the spoken word. The smartphone then used the phone-function as an excuse to combine it with a virtually unlimited number of additional functions. All of a sudden, a phone is not (just) a phone. What will "phones" look like in the post-touchpad era? Will they become ever smarter? Is the incremental approach to innovation and problem solving inevitable or is there a shortcut?

Borrowings can be a kind of shortcut.

The Borrowings. Knowledge and skills specific to a domain are inside sources. Borrowings come from outside sources, i.e., other domains. They are things that are ***noticed when needed.***[9] From whence then, come the borrowings? Strategies, contingencies, collaborations, bricolage, and the initial state of an innovation problem.

Strategies

Strategies can be defined in two ways. In the first, they are specific, alternate paths to the same goal. For example, there are six different strategies for solving single-digit addition problems. Strategies of this kind are domain-specific. They are basic, not borrowed.

In the second, a strategy is a general procedure for solving problems across domains. They are basic to problem solving, but are borrowed by other domains. The paired constraint model is this kind of strategy: noticed when needed, borrowed not basic.

Couldn't you say that, in a very general sense, all borrowings are strategic?

Yes, and in the most general sense, borrowing itself is a strategy.

Contingencies

Pat, it is here where the work I have done on thematic thinking comes in handy (Froehlich, Gibbert, & Hoegel, 2014). *That is, we see the "need" when there is a*

[9]This idea is also covered in *Digression I: The Problem Solvers.*

spatio-temporal contingency in some way. For instance, the phone was combined with the camera, since the phone keeps the camera "handy" (in fact, the German term for mobile phone is "handy"!). Consider the sofa-bed, a truly magnificent innovation (dating back to 1899) which can be found in most house-holds. As with the camera-phone, the innovation here comes from borrowing close resources and combining them in synergistic ways. By "synergistic" I mean in ways that go beyond the functions each individual product could do on its own. By keeping the camera "handy" in the phone, we are transformed into real-time reporter/journalists in a way even George Orwell would have thought implausible. In terms of the sofa-bed, what makes this innovation so magnificent is its power to transform your living room into a kind of on-demand bedroom (or the other way round). Which other piece of furniture has the power to transform the very essence of the room it is surrounded with?

Incidentally, the sofa-bed, invented by Leonard C. Bailey, an African-American born into poverty, is a neat illustration of domain-specific deliberate innovation. If you are born into poverty, you probably will have spent many nights sleeping on a sofa (or using somebody's bed as your sofa), so merging the two makes sense. If you are the son of a goldsmith/engraver (to come back to Gutenberg), probably the last material you think of for making moveable type is wood (the technology for printing before the movable type consisted in carving an entire printing page into wood). Engraving every single moveable metal-type would have been even more expensive than carving typeface into wood, though... How does your father produce cheaper engravings? By engraving only the mold and then casting molten metal into that mold. But then, precious metals are too, well, precious to turn them into moveable type. So which other metals can be used and easily molten? A train of thought about contingencies like these presumably led young Johannes to the idea of using a lead (easily molten over the household fire, something most kids will have observed their fathers doing when casting bullets for hunting), reinforced with a bit of tin and antimony for extra durability.

The printing press on its own would not have freed (liberated...) the individual letter head from its wooden printing board where it was firmly fixed in sentences and paragraphs. Economically-produced die-cast letterheads, in combination with the printing technology did, however. Such contingencies lead to synergistic combinations, which outsoar the domain-specific provinciality from which they emerge.

Michael, let me expand a bit on contingency. In my understanding, a contingency is a relationship in which one thing is dependent on another. In your examples, noticing is dependent on three contingencies—one is need (there is no noticing without need), the second is chance (the occasion on which the thing needed is noticed), the third is knowledge (recognizing that the thing noticed *is* the thing needed). So, for noticing to happen we have (1) need, (2) chance, and (3) knowledge.

I want to add two things on which both noticing and knowledge are dependent: plasticity and variability. Plasticity refers to the ease with which the brain adapts to input from the world (Nelson, 1999). Adapting here means rewiring, physical/tangible rewiring that creates new and expands existing neural networks. Plasticity induces noticing, noticing further expands the networks. What we call talents and gifts are based on increased plasticity in particular parts of the brain. For example, mathematicians notice patterns,[10] composers and musicians notice sounds, painters notice colors.

While plasticity is pre-wired, variability is acquired.

My earlier research showed that learning involves both how to do something (the skill) and how differently to continue doing it (the variability level). The level is determined by task difficulty early in learning. If success/mastery depends on trying many things, high variability will be rewarded. If rewarded, it will be repeated, becoming habitual in (and only in) the training domain. This is what makes all of us more or less habitually/predictably variable in different areas, say writing or proto-typing or cooking. There are other contingencies. High variability levels facilitate skill acquisition and transfer. They also increase sensitivity to (i.e., *noticing*) changes in condition, and subsequent alteration and adaptation. I could go on—the contingencies (including those between variability and plasticity) are continuous—but I should switch to collaborations (Stokes, Lai, Holtz, Rigsbee, & Cherrick, 2008).

Collaborations

Some innovators, experts in multiple domains, collaborate with themselves. I'm thinking as usual, about artistic innovations, about Alexander Calder's mobile. The mobile was the product of his initial training as an engineer (outside source) and his later development as a sculptor (inside source). Paul Klee had to choose between careers in music or art. Art won, but music left its mark on both the vocabulary (the Canon of Tonality parallels the cycle of fifths in classical music) and the rhythms (the directional patterns in his grids represent tempi like staccato and legato) of his paintings. For my math program, I had two things at hand: the Japanese count that I learned working in Tokyo and the design skills I acquired at Pratt.

Many more innovators collaborate with outside sources, borrowing from other experts' tool boxes. These collaborations can be direct, person to person interactions, like those between theorists and engineers at (legendary) Bell Labs, between design teams at (becoming legendary) IDEO, or between engineers and data analysts at Vestas Wind Systems, a Danish wind power company, where turbines are designed and built to match wind conditions mathematically modeled for the exact place where each turbine is placed.[11]

---

[10] In answer to my question "How did you become a mathematician?" a mathematician friend (Andrew Sanfratello) told me that, from the time he was very young, he always noticed patterns, he only noticed patterns, all he was interested in were patterns. Every time I see Andrew, I ask him to tell me something new about numbers and he does and the patterns are amazing!

[11] This piece of information was "noticed" by Pat, in the New York Times business section, on the day she wrote this section on collaborating.

Collaborations can also be indirect. *Gutenberg noticed the screw press used in wine making, thought in terms of metal (not wood), and saw the link between engraving and die-casting. Being the son of a goldsmith, he probably thought in terms of small artifacts made of metal, so the idea of die-casting entire printing pages must have seemed outright grotesque.* Impressionism happened (light breaking up on things vs. light reflected from things) when Monet noticed Chevreul's color wheel. Netflix happened when Reed Hastings noticed memberships (vs fees) from his gym, mail order (vs. in-store) from Amazon, and—then recently introduced from Japan—DVDs (vs. videocassettes). Some indirect collaborations are intergenerational. I quote Isaacson on this: "The best innovators were those who understood the trajectory of technological change and took the baton from innovators who preceded them. Steve Jobs built on the work of Doug Engelbart, who built on J. C. R. Licklider and Vannevar Bish" (Isaacson, 2014, p. 480).

Bricolage

Pat, the way I see it is that resource constraints make us notice stuff that is right in front of our eyes, and which can hold the key to an innovation solution, but which is kind of eclipsed by having enough money to invent around the problem. There is a nice case on the military jet engine here. As I wrote in a paper with my colleague Phil Scranton for "Management and Organizational History" (Gibbert & Scranton, 2009), *the challenge with jet propulsion is quite simple: the hotter the engine runs the higher its fuel efficiency and the greater its thrust. If the whole machine gets too hot, it unfortunately blows up. At the end of the Second World War, the Allies competed against the Germans in the development of what was seen (even more so than the atomic bomb) as THE "winning weapon" to end the war. On the Allied side, the US government prioritized the jet engine over the atomic bomb to the extent that the funding the military jet engine was double that of the Manhattan Project. Now, how would you go about solving the fundamental challenge in jet propulsion if you had all the money you needed? The American engineers focused on creating more heat resistant alloys (which needed a lot of hard-to-come-by precious metals) in an effort to turn the entire engine less susceptible to crack down when it ran at full speed. So, the project looked something like this: cook up a new alloy, produce the engine, run it at full speed, and watch it blow up. There was a much better solution to the problem.... (to be continued).*

The Initial State

I originally had this under collaboration, but I think it's important enough to warrant its own sub-head. What's so important? Realizing that your most critical collaborator is the initial state of your innovation problem—the specific style, product, or process you aim to re-specify, refine, or replace. In many cases, obviously those of necessity and less obviously those of style,[12] you do not choose your initial state. If you can, choose it carefully, strategically.

Remember (like the Pressman), it supplies the items in your preclude column, without which you have no place to start (inventing the Walkman).

Other Things to Remember

- Innovation is a problem-solving process.
- You can only problem solve inside the tool box of your expertise.
- You can make your tool box bigger.
- Paired constraints can make it bigger.
- You should make it bigger because....
- The innovative-ness of your solution depends *entirely* on its contents.

References

Ahuja, G., & Katila, R. (2004). Where do resources come from? The role of idiosyncratic situations. *Strategic Management Journal, 25,* 887–907.

Baker, T., & Nelson, R. E. (2005). Creating something from nothing: Resource construction through entrepreneurial bricolage. *Administrative Science Quarterly, 50,* 329–366.

Cunha, M. P., Cunha, J. V., & Kamoche, K. (1999). Organizational improvisation: what, when, how and why. *International Journal of Management Reviews, 1,* 299–341.

Ecco, U. (1994). *Reflections on the name of the rose*. London: Secker & Warburg.

Froehlich, J., Gibbert, M., & Hoegel, M. (2014). *Thematic thinking*. Upper Saddle River, New Jersey: Financial Times Press.

Gibbert, M., Hoegel, M., & Valikangas, L. (2007). In praise of resource constraints. *MIT Sloan Management Review, 48,* 15–17.

Gibbert, M., & Scranton, P. (2009). Constraints as sources of radical innovation? Insights from Jet Propulsion development. *Management & Organizational History, 4*(4), 01–15.

Isaacson, W. (2014). *The innovators: How a group of hackers, geeks, and geniuses created the digital revolution*. New York: Simon & Schuster.

Katila, R., & Shane, S. (2005). When does lack of resources make new firms innovative? *Academy of Management Journal, 48,* 814–829.

Miner, A. S., Bassoff, P., & Moorman, C. (2001). Organizational improvisation and learning: A field study. *Administrative Science Quarterly, 46,* 304–337.

Motherwell, R. (1992). *The collected writings of Robert Motherwell*. Oxford: Oxford Unversity Press.

Nelson, C. A. (1999). Neural plasticity and human development. *Current Directions in Psychological Science, 8,* 42–45.

---

[12]Major shifts in painting styles are usually a direct response to a currently dominant style. A good example is Pop Art, which began with a list contrasting features of Abstract Expressionism (painterly, emotional, improvisational) and their opposites (mechanical, ironic, planned). The opposites became the substitutes that characterized the new style (Stokes, 2009, 2016).

Rivers, L. (1987, March 31). *Improvisation and the creative process in jazz and the visual arts*. Presentation given at Barnard College, Columbia University, NY.
Stokes, P. D. (2006). *Creativity from constraints: The psychology of breakthrough*. NY: Springer.
Stokes, P. D. (2009). Using constraints to create novelty: A case study. *Psychology of Aesthetics, Creativity and the Arts, 3,* 174–180.
Stokes, P. D. (2016). *Creativity from constraints in the performing arts*. New York: Stokes.
Stokes, P. D., Lai, B., Holtz, D., Rigsbee, E., & Cherrick, D. (2008). Effects of practice on variability. Effects of variability on transfer. *Journal of Experimental Psychology: Human Perception and Performance, 34,* 640–659.
Weiss, M., Hoegl, M., & Gibbert, M. (2014). Making virtue of necessity: The role of team climate for innovation in resource-constrained innovation projects. *The Journal of Product Innovation Management, 32,* 278–291.

Chapter 3
What's Your Problem? An Innovation Typology

Innovations can emerge if they resolve existing problems in new ways... alternatively, they can emerge if they solve new problems.
Ahuja and Katila

Some problems are more generative than others.
Michael

Some paths are more generative than others.
Pat

This chapter is our attempt to categorize innovation problems on the basis of novelty in path and/or goal. I have argued (and continue to argue) that a new solution path is necessary—it *is* the innovation. Ahuja and Katila (2004) agree with me: their categories are problems of scarcity and opportunity, both of which require new paths; only opportunity adds a new goal. Michael suggested another possibility which I questioned: *an innovation with a new goal, but not a new path.* Really Michael?

Really Pat. And also one initiated with neither a new path nor a new goal.

Initiated helps. Do you mean that an innovation can ***start*** with Same Path/Same Goal?

Exactly! You just said something very important here, Pat. You are pointing to a transition phase from one category in the matrix to another. This is really key. Some innovation problems start out as one type and then are re-cast into another.

To examine possible transitions, as well as classic categories, Michael suggested a two-by-two matrix.

P. D. Stokes and M. Gibbert, *Using Paired Constraints to Solve The Innovation Problem*, https://doi.org/10.1007/978-3-030-25771-2_3

Table 3.1 Goal by path matrix

PATH		
GOAL	Same	New
Same	**I** **Same Path/Same Goal** Transition → New Path/Same Goal	**II** **New Path/Same Goal** Scarcity
New	**III** **Same Path/New Goal** Transition → New Path/New Goal	**IV** **New Path/New Goal** Opportunity Aspiration

The Matrix

In the matrix, there are two kinds of Path (Same and New) and two sorts of Goal (Same and New), which give us the following categories. The transition categories are indicated as such.

- *Category I: Same Path/Same Goal (transition),*
- *Category II: New Path/Same Goal,*
- *Category III: Same Path/New Goal (transition),*
- *Category IV: New Path/New Goal* (Table 3.1).

Category I: Same Path/Same Goal → New Path/Same Goal

This is our first transition category. In it, we find innovations that ***start*** with Same Path/Same Goal.

The example that came quickest to (Pat's) mind was MOOCs or Massive Open Online Courses. MOOCs make college courses traditionally (and still) taught in classrooms (for tuition) online and free. The goal is the same: educate students. The sameness in the path is the content: the course itself taught by a college professor. Altering the path, putting the course online and without charge, makes it new and innovative. In Chap. 4, I'll write about the development of one early MOOC, *Courserva.*

I'm guessing that Uber and Hasty also belong here. The Economist (Jan. 3, 2015, p. 9) wrote about gig-economies that "supply luxuries once reserved for the rich." Uber supplies chauffeurs. Hasty sends housekeepers. That also makes part of each path the same. The way those services are delivered (on-call rather than on-staff) is what occasions the detour to new goal/new path. In both examples, the product (the course and the service) and the goal (learning and getting somewhere) are the same; the difference is in the cost and the delivery.

Vintage

A recently re-invigorated product category that ***stays*** in Same Path/Same Goal is vintage. Think of high-end classic watches (self-winding) and boats (wooden sailboats and yachts), also more affordable turntables and vinyl records. For example, in time for Christmas 2017, three archival collections of Thelonious Monk were released: two only on LPs and only one on LPs and CDs.

In a sense, self-winding and vinyl preclude digital and promote analogue. But since neither paths nor goals are new, there is no transition and no innovation.

Category II: New Path/Same Goal

This is the category on which we all (Michael, me, Ahuja, and Katila) agree. There are two catalysts for innovation here. One is what Michael called the "mother of innovation"—necessity: a new path is required. The other is desirability.

Necessity

Here, resource constraints on materials or skills preclude specific steps along an existing solution path. Substitutions are required to resolve the problem, i.e., to satisfy the same goal criterion. An example is my pesto problem, where the absence of some ingredients led to a different path, albeit with the same goal (making pesto).

Another illustrative example is needing to water geraniums on the windowsill with a tea pot (because we are without a watering can). The creative spark here comes from leaving the design stance (the goal the object was designed to achieve) and seeing its applicability to solve the problem at hand. Again, the keyword is "bricolage," using objects and parts of solution paths to achieve a purpose that they were not built to achieve.

Desirability

Choice is desirable, at least to a degree.[1] Choice depends on permeability—alternate ways to arrange or rearrange a set of elements. Think about the permutations in fabric (i.e., wallpaper) patterns, in Mondrian's grid paintings, and in too many

[1]Choice, as shown by Schwartz (2004), is also stressful. Too many options can lead to none chosen.

varieties of breakfast cereal or face cream. Permeability also occurs in mathematical solutions and in musical performance. There are multiple ways to break a number down into its prime factors.[2] There's something my piano teacher called a "musical intention," and it's how a skilled pianist intentionally varies his playing. One element that can be varied is tempo. In my book of sheet music for Bach's two- and three-part inventions, there is a page listing the tempi at which the 15 Sinfonias (the three-part inventions) were recorded by 13 well-known pianists. The tempi were radically different. On 11 of the 13, Glenn Gould, not surprisingly, played faster than the others: same goal criterion (perform Bach), new (because permeable) path.

Sorry, not sure I follow here. If the tempi were different, doesn't this mean the path is different? Yes. This is New Path/Same Goal (different tempi/same Bach composition).

Category III: Same Path/New Goal → New Path/New Goal

This is where (I am quoting you Michael) the innovator thinks about "what else to make with what he has." Making something "else" could include re-positioning/promoting your product to a different audience. This precludes one audience/product benefit and promotes another: thus, a new path.

Let me use one of your examples Michael, the Post-It Note.[3] Here, the path produced a new glue that wouldn't quite stick. Someone at 3M *noticed* an opportunity—using the not-so-sticky glue to launch a new product. Now the Post-It Note moves over to Category IV: New Path/New Goal, or least I think so.

I agree. Post-It Notes are good example of the path clearly changing (applying the glue to small bits of paper, rather than selling it on its own), and the goal most definitely changing (the whole point about glue is that it is supposed to "stick"). So, it belongs in the transition to New Path/New Goal category as well.

Category IV: New Path/New Goal

This category also has two catalysts. One is innovation's "fairy godmother"—opportunity.[4] In cases of opportunity, something is noticed. The other is aspiration. In aspiration, something is next. Let me elaborate.

[2]For example, 12 can be initially broken down into 2×6 or 4×3. Six breaks down into 2×3; 4 into 2×2. Either path produces the same prime factors: 2, 2, and 3.

[3]See Gibbert and Valikangas (2004).

[4]Opportunity was Possibility 3 in Chap. 1.

Opportunity: Something is Noticed

I'm thinking of Chevreul's color wheel, noticed by Monet; of the Italian espresso bar noticed by Schultz (Starbucks); Amazon's delivery and his gym's membership-payment systems noticed by Hastings (Netflix); the Japanese count noticed by me (I'll get to that in Chap. 5). Columbia's noticing that silence (three-minutes between songs) was saleable also belongs here, as does 3M's noticing that the not-so-sticky glue that made the Post-It Note possible.

I'm also thinking of what I've called "generativity."[5] Here, the initial noticing leads to further, more highly focused noticing, which further innovation by fostering sub-goals and still newer paths. Monet is, of course, the exemplar.

Aspiration: Something is Next

This is how domains are expanded. The next "big thing" is always based on a current big thing. Why? Because the current big thing provides that the initial state and the preclude column from which innovations—in animation (Pixar's commuter generated), transportation (Tesla's electric car), publication (Amazon's print-on-demand), and painting (Pop)—emerge, substitution-by-substitution. As we'll see, generativity also comes into play here. The next big thing, on occasion, occasions noticing opportunities for development.

In Chap. 7, I'll develop problem spaces for Pixar, Starbucks, a well-known Pop painter, and an innovative math program, *Only the NUMBERS Count*. Michael will tell us about sous-vide and the BlaBla Car.

The Take-Away: Four Problems/Two Transitions

I think we've made our matrix more complicated—but also more useful for structuring your innovation problem—by characterizing the transitions or detours within it.

To summarize, the paired constraint model specifies that an innovation requires a new solution path. No transitions are required in classic cases of scarcity (New Path/Same Goal) or opportunity (New Path/New Goal). Transitions are required in cases (like MOOCs) that start in Same Path/Same Goal or (like Post-It Notes) that begin in Same Path/New Goal. Thinking in terms of paired constraints, innovation precludes two problem types (Same Path/Same Goal and Same Path/New Goal) and promotes two substitutes (New Path/Same Goal and New Path/New Goal).

We elaborate on all in the following chapters.

---

[5]Generativity was Possibility 4.

References

Ahuja, G., & Katila, R. (2004). Where do resources come from? The role of idiosyncratic situations. *Strategic Management Journal, 25,* 887–907.

Gibbert, M., & Valikangas, L. (2004). Boundaries and Innovation: Introduction to the special issue. *Long Range Planning, 37,* 495–504.

Schwartz, B. (2004). *The paradox of choice: Why less is more.* New York: HarperPerennial.

Chapter 4
Category I: Same Path/Same Goal to New Path/Same Goal

Routines need to be redesigned for new contexts.
Gupta, Hoopes & Knott

Some innovation problems start as one type and then are re-cast into another.
Michael

The goal can remain the same; the innovation can involve only the delivery.
Pat

This chapter covers transitions from Same Path/Same Goal to New Path/Same Goal. Our focus here is on familiar products with innovative delivery systems. Our examples include a luxury (online rather than on-staff) once limited to the rich and three classes (two hands-on and one online). The goal (the specific product or service) is the same. The difference (and the innovation) is in the delivery (specified by the novel solution path).

Transition I: Calling the Chauffeur

I am sure that some readers will quibble with my calling an Uber driver a chauffeur, but there is a similarity in the convenience of having a car and driver not just at hand, but in hand. What Uber's app promises (besides not watching for an empty cab or waiting for a car service to let you know they've arrived) is instant gratification. You call, and voila, you can watch your chauffeur (whose name you know) arriving! Does this put Uber in the New Path/New Goal category? I don't think so. The product (car and driver for hire) is the same as that offered by cabs and car services, the difference is the delivery. The difference is highly innovative, but the goal is the same. This means that car services and cabs are sources for the preclude column in Uber's problem space. Let's look at the substitutions (Table 4.1).

P. D. Stokes and M. Gibbert, *Using Paired Constraints to Solve The Innovation Problem*, https://doi.org/10.1007/978-3-030-25771-2_4

Table 4.1 Uber et al

Initial state: Hiring a car and driver			
Search space	**Constraint pairs**		
	Preclude	→	**Promote**
	Cabs		
	Hail on street		Hail on app
	Uncertain arrival		Certain arrival (see car on way)
	Metered fare		Estimated fare
	Car service		
	Call in advance		Call when ready
	Set fare		Dynamic pricing (determined by demand)
	Company cars		Driver-owned cars
	Both 1 and 2		
	Private ride		Shared rides
	No ratings		Ratings of drivers and riders
Goal state: Hiring a car and driver			

The differences are, as I said before, in the delivery: hailing your car and driver on your app when you're ready to go; knowing how your driver was rated by other riders, where your car and driver are, and also what your fare will be. Once you get in the car, the differences disappear. The end product (car and driver) is the same whether you hailed a cab, called a car service, or tapped the Uber app.

Pat, have you heard of BlaBla Car? It puts car and driver in a whole new category.

Tell us more please Michael.

Of course, but in the chapter on New Path/New Goal.

Transition II: Cooking School at Home

A young woman I know selects (from a weekly menu) 3 Family Plan meals from Blue Apron. For the week of January 22, 2018, she might have chosen these meals:

Steaks and warm lemon salsa verde with roasted broccoli and sweet potatoes.
Orange and brown butter tilapia with garlic rice and collard greens.
Pesto and goat cheese fusilli with swiss chard.

Each would serve 4 people (the size of her family), arrive in a refrigerated box, and come with pre-measured ingredients and step-by-step instructions with pictures from unpacking to plating, plus online videos for techniques like chopping herbs. The menus change each week. They are designed to replace repetition (most of us have small repertoires of daily dishes) with interesting and, importantly, *failure-proof* meals. Instead of a chef, Blue Ribbon provides (like a cooking school) a series of complete lessons in becoming one.

Table 4.2 Blue Apron

Initial state: Home-cooked meal			
Search space	**Constraint pairs**		
	Preclude	→	**Promote**
	Repetitive, unappetizing		Interesting, appetizing
	Plan meals		Select meals from menu
	Shop for ingredients		Delivered ingredients
	Lack of expertise		Cooking lessons
	Measure		Exact amounts
Goal state: Home-cooked meal			

Here is my problem space for Blue Apron. Notice that (like Uber's) the initial and goal states are the same (a home-cooked meal). Also, like Uber's, the difference is in the delivery—more specifically here, in what is delivered (Table 4.2).

My young friend is delighted (and also, she says, delightfully surprised) with the meals she now makes. My question was, now that she knows how to make lots of great meals, is she still using Blue Ribbon? Her answer was yes. It's too convenient and too good to give up.

Pat, there are several such platforms also in Europe, all of which are run by startups. The marketing Prof in me makes me wonder why none of the major supermarket chains have entered this new business, especially since the ***home-cooked meal*** *initial state is their core business. Mind you, many supermarkets have started to have small sections dedicated to pre-cooked or otherwise easy-to-prepare (cold cuts, salads, etc.) meals (typically for one-person) complete with half bottles of wine (I am talking Italy here...), and crispy bread. The idea of bringing together items that are usually in different aisles in the supermarket is what the psychologist I mentioned earlier,* Barsalou (1983), *calls "goal-based categories." The psychology-textbook example for a goal-based category is "things to take out of the house in case of fire." And this includes anything from valuables (be the small such as jewelry or large-but-portable such as paintings), to things that may be less valuable from a monetary perspective but to which we are attached such as photographs, diaries, etc. Goal-based categories are different from classical, taxonomic categories, according to which supermarkets are typically organized (think different types of pasta, rather than pasta and canned tomatoes), but which have you gallivanting through the supermarket aisles in search of the constituents for your supper (those canned tomatoes are usually far from dried pasta, even in Italy).*

So, why don't supermarkets pick the Blue Ribbon idea up? Most even have home delivery these days, and online ordering makes it easy for them to suggest different combinations (a.k.a. different meals). They already have their customer base (Blue Ribbon didn't when it started), and even know who is regularly ordering online. Why not offer these people personalized suggestions (the supermarket can tell from the order history who is vegetarian, has gluten allergies, etc.)? All it takes is a bit of New Path/Same Goal thinking…

Update!

At least one supermarket (Albertson's) and one super chain (Walmart) were thinking your way Michael. Each now offers meal kits, but differently form the originals. No subscriptions. No choosing dinner days in advance. Instead, the meal kits are sold in-store on a day to day basis. Is it working? Must be. Blue Apron plans to sell its kits in this year.

Where did I find this? In the *Economist*, April 14, 2018, months after we wrote this section.

Transition III: College Classes at Home

Most homeschooling involves grammar school students. A more advanced option is distance learning, which started via the mail as correspondence courses and continues online as MOCCs or Massive Open Online Courses. Massive refers to class size (unlimited); open to participation restrictions (none); online to delivery method; and courses to content (with or without certifications or degrees). Coursera, launched in 2012 by two Stanford University computer science professors, now offers over 2000 courses to more than 28 million registered users.

To write this section, I signed up for a free course on Design Thinking (which is, of course, related to innovation). What did I learn (about Design Thinking and about the substitutions made to move in-class to online learning)? Not much.

I was disappointed. Design Thinking—to someone who went to Pratt and was a creative group head in advertising—is nothing new. However, it would be new (and too complicated be useful) to someone without my particular expertise. So, let me focus instead on format rather than content. The course was structured like a typical college class, with all the downsides of a typical college class.[1] In other words, there were virtually no substitutions in presentation (which was very static), only in delivery (Table 4.3).

Michael, who is a full-time professor, has problems with MOOCs that are bigger than mine. His focus is on the industrialization of tertiary education. What this means, he told me, is that *only a couple of star-teachers will get/keep their jobs and the coordination of these courses (someone needs to do the grading) is done by assistants. It also deprives the students of the critical social interactions (and networking opportunities) a real campus experience provides.*

---

[1]Students at Columbia have repeatedly told me that it doesn't pay to attend lectures where professors post their power-point presentations online. In class, many simply read the power point. What is the point of attending?

Table 4.3 Coursera

Initial state: College-level class			
Search space	**Constraint pairs**		
Retain	**Preclude**	→	**Promote**
Lecture	On-campus		Online
Power point	High tuition		Free for no-credit/fee for certification
Visiting experts	Fixed time and place		Flexible time and place
Discussion groups			
Goal state: College-level class			

I think there's a more basic problem on-campus: the reason there are so few star-teachers is because no one teaches professors how to teach. (Acting classes would be a good start). Importantly, given the pre-packaged presentation of material in most classes, Michael told me about a different delivery system (which includes lots of interaction). It's called the Flipped Classroom.

Transition IV: Flipping the Classroom

Let me try a different example for changing the path to achieve the same goal: the flipped classroom. It goes back to a Harvard colleague **Eric Mazur** *who outlined the basic idea in his book* ***Peer Instruction: A User's Manual*** (1997). *The flipped classroom is called "flipped" because that's what it does: it inverts the tasks that are done at home and the tasks that are done in class. Typically, you have students read the relevant book chapter or other readings in preparation for class. Class time is then spent on regurgitating what the students should already know, using tons of power-point slides. It comes as no surprise that students have little motivation to actually prepare for class, since the contents of the average textbook are too much anyway to regurgitate during the average class time, so the professor needs to prioritize. Why study the whole chapter when the professor then only focuses on what he or she thinks is the most important stuff?*

The flipped classroom (in its classical variant) has students watch pre-recorded lessons (similarly to MOOCs, except that they are not "massive," i.e., unlimited, and are typically the professor's own recordings), and then class time is spent on applying the concepts/formulas/ideas to actual cases, examples, and problems. The result is a much more hands-on approach to learning, where students cannot "escape" (unlike traditional classrooms, where they sit behind their laptops or tables and engage in all sorts of things that tend to be only vaguely related to what the professor is talking about). It puts the student where he or she should be: at the center of attention (dethroning the professor). So, time at home is dedicated to good old-fashioned studying and in class, there is a reality check that this material has actually sunk in. Performance measures can take many forms, from regular individual assignments to group-based projects.

In my own courses,[2] *I use a blog-based system where students work in groups during class time on concrete problems (some of which I give them and some of which they can pick themselves) and where they are asked to rigorously apply the studied concepts. At the end of class, all groups need to upload their material on the blog. This gives the professor an incredibly powerful way of checking the sophistication of their understanding. The students get feedback the day after in written form (via comments on their blog posts) and are encouraged to comment on their peers' blog posts (and their doing so is part of the participation grade they get in the end). I've been in this job for over 15 years and have never before had the feeling of being SO close to my students, their understanding, and my own successes and mishaps in attempting to relay the material to them.*

Michael, what I especially like about your flipped classroom is that students are "learning by doing," something long advocated by Herb Simon (Anzai & Simon, 1979). My question is this: given the methodology, are these classes small, seminars and/or graduate level?

Not really, I also teach this way to undergraduates in 70–80 student classes and it works just fine. It just requires more assistants for the group work.

References

Anzai, Y., & Simon, H. A. (1979). The theory of learning by doing. *Psychological Science, 86*, 124–140.

Barsalou, L. W. (1983). Ad hoc categories. *Memory & Cognition, 11*, 211–227.

Mazur, E. (1997). *Peer instruction: A user's manual*. New York: Pearson.

[2]See, for instance https://consumerbehaviorexamples.wordpress.com/.

Chapter 5
Category II: New Path/Same Goal

Fashion changes but style endures.
Coco Chanel

If necessity is the mother of innovation, who is the father?
Michael

Permeability makes multiple paths possible.
Pat

Unlike Uber or Blue Ribbon, which are luxuries, this chapter examines innovations that involve scarcity or, more generally, necessity. Michael's (classic) quote points to substitutions that are required; Coco Chanel's to those required and desired. Permeability, alternate ways to arrange or rearrange a set of elements, underlies both.

Substitutions Required

Coca-Cola is a classic example. The original formula included coca leaves (for cocaine), wine (for flavor), and kola nuts (for caffeine). *Necessity* arrived twice—first prohibition precluded the flavor supplied by the wine, requiring a sugary syrup substitute; second, a new federal tax on drugs, precluded the cocaine, requiring a new, still-secret formula.

My pesto problem also belongs here. *Scarcity* was the source of the innovation here. No basil, no pine nuts, no pesto. Substitutions: parsley, walnuts. Solution: a pesto of a different sort for (in my example) the same simple supper.

Pat, let me elaborate on your pesto problem. It belongs to a category of required substitutions in which one element in a configuration is replaced by another of the same configuration. In the pesto example, for instance, if you run out of pine nuts and walnuts, in fact if you run out of nuts altogether, you simply substitute with another white-ish powder which gibes the pesto body, amalgamation, and taste: parmesan cheese (or indeed, pecorino). David Mazursky from Hebrew University Jerusalem

P. D. Stokes and M. Gibbert, *Using Paired Constraints to Solve The Innovation Problem*, https://doi.org/10.1007/978-3-030-25771-2_5

called this innovation principle "replacement" and hailed it as a veritable "innovation template" (Gibbert, Goldenberg, Horowitz, Levay, & Mazursky, 2006). *The basic principle is standard in industrial design, where you want to minimize, not multiply, the parts of a product. David's classic example is a chair for a child's table. The chair is attached to the table itself, using the table's legs as its own.*

Other examples, please.

Margarine, Etc

Wartime examples abound. When there is no butter, we make vegetable spread and call it margarine. When the war ends, and there is more than enough butter, we hire marketing and nutrition professors to tell us that margarine is actually better for you than butter. The problem with many of these substitution-required solutions is that they substitute one evil for another. Where once you worried about clogged up arteries (from the milk fat), you are now faced with the risk of cancer (from the chemicals that make the vegetable oils spreadable). A prime example in the USA is corn syrup, much cheaper than (cane or beet) sugar (again financial constraints leading to a required substitution), but also more conducive to diabetes.

Skype

In the early days of Internet telephony (then-called "Voice over Internet Protocol"), there were several solutions to connect the computer (then the only device connected to the Internet) with a receiver necessary to make "phone calls." YaYa (a now defunct service) used an existing component of the configuration, your telephone (a landline) as the call-back device. Basically, you dialed via YaYa's platform and the computer rang your phone and you made cheaper calls.

It took Skype (and computer manufacturers) some time to work the substitutions out. Do you remember when "skyping" (calling someone) was a neologism and we had to rely on elaborate headsets (which made us look a little like Britney Spears in her early videos) to use the Luxembourgian non-telephone substitute? Skype even used this to launch their own line of Skype-branded headsets and other gear. Then Sony came out with a cleverer solution: their VIO line included a telephone-mouse hybrid, which let you use your mouse as a telephone receiver. Problem was picking up and taking into your mouse made it difficult to use the mouse for what it was designed to do (unless you had a laptop where a back-up mouse was integrated into your keypad). An even bigger problem was that people (including your co-author) mistakenly picked up the mouse when the "cell-phone" (which tended to lie next to the mouse-phone or shall we say the phone-mouse) rang—or even used the cell phone as a mouse.

I actually wonder why it took industrial designers so long to substitute (David would say "replace") these rather cumbersome devices with a microphone integrated into the computer itself.

Perhaps because the substitution was not required but desired.

Or at least "desirable." Note that Skype came of age when "cell" phones became "smart" (i.e., connected to the Internet) and Skype became a direct competitor to your telephony provider. I remember inviting a Vodaphone marketing manager to speak to my students once and he proudly presented an Internet flat-rate for all kinds of "strategic partners" of Vodaphone, such as Amazon, and eBay. When I asked him if that included Skype, I got a grim look. While having a flat rate for Skype was clearly desirable (and even required) from the consumer perspective, it was much less so from the provider's.

Which bring us to the topic of the next section.

Substitutions Required and Desired

When Michael told me that he wanted to write about sous-vide here, I thought "wrong" that's a completely new category. Turns out that in its first incarnation, it was not a new category, but rather a technique for reheating food (required) without reducing its flavor (desired).

Sous-Vide I: "Shorting" the Flavor

The first time I came across sous-vide cooking was during my apprenticeship as a chef in Wiesbaden. The idea really is quite simple, you put whatever you want to cook into a vacuum pack and toss it in water of the appropriate temperature (which by definition cannot exceed 100 degrees and indeed is typically much lower) until it's done. Sous-vide "cooking" in professional kitchens tends to be a misnomer, since the "cooking" typically is not about cooking a raw ingredient, but re-heating an already cooked meal as efficiently as possible. Let me explain. Say you have pieces of meat that require slow-cooking in liquid (the jargon is "braising"). Ideally, you want to use as little cooking liquid as possible to concentrate flavor (the jargon here is keeping it "short"). Keeping the cooking liquid short is easy for dishes such as goulash, blanquette, fricassee, and, say, ragout, since the meat pieces are small and easily cooked in a liquid that barely covers them. Keeping the liquid short is much more difficult for larger pieces, especially pieces that still have the bones in and have not been cut up in small cubes, e.g., lamb shanks (in ragouts, the meat is still on the bone, but in smaller pieces). An individual lamb shank, for instance, is difficult to re-heat unless covered with an adequate amount of liquid (ideally the tasty gravy it was cooked in for hours). Sous-vide here solves the problem by putting a small amount of the gravy together with the shank into a vacuum pack and slowly re-heat the whole thing once an order comes in (the vacuum packing also makes it much easier to store—even freeze—the individual portion). The same is true for pot-au-feu, or as we call it in Italy, bollito misto, or, say, Irish Stew.

I have a question: Why not just micro-wave?

Pat, you do make me laugh at times. You Americans chuck just about everything into the microwave, even popcorn. Mind you, this has led to one of the most useful product innovations in the microwave category in decades, the popcorn button. I wonder when the fois-gras button will pop up (quite literally). But seriously, the problem with microwaving is the uneven distribution of the microwaves on the inside of the meaty-bony bits. You wind up overheating some bits (e.g., the carrot pieces in the pot-au-feu) and under heating others (the meat closest to the bone of the chicken drumstick). Another problem with the microwave, especially in a professional kitchen, are space constraints: you can only heat one (or maybe two) portions in one go. The bain-marie variant is easily scaled up by putting a bigger pot on the stove.

Fashion I: Interpretation

An obvious domain in this category is fashion. To be fashionable means to be *au courant*, to be "in style," in the current style. The goal (to be fashionable) remains the same. The solution path (the newest style) changes. The question that interests me is: how do "houses" (e.g., house of Dior) keep their newest collections recognizably theirs? This changes the goal somewhat:

Goal State: To be fashionable (a new collection)

Sub-goal: To be recognizable (a valued "label")

The recognition factor is desired both by the designers and by the clientele. After all, who would pay for a Dior that did not look like a Dior? So, how do the designers do it? Interpretation or translation is one way. Here, there are two things that can be interpreted. One is the recognizable look of a fashion house, say Chanel or Dior. The other is a look borrowed from an era, perhaps popularized in a film or show. Think of Ralph Lauren's safari season, adapted from *Out of Africa*.

What makes a Chanel a Chanel? This was the question Karl Lagerfeld (head of design at Chanel until 2019) had to continuously answer. His job was to keep the classic Chanel brand recognizable while making the clothes themselves contemporary. What he retained (always in part) were the classic elements: the little black dress, the plaid boxy jacket, the quilted metal-chained handbag, the oversized fake pearls and peonies, the capped toe shoe, Number 5 perfume, and of course, the logo. The elements were updated to fit current stylistic, functional, and material constraints. Lagerfeld was fortunate. Most women would love to wear the classic black Chanel dress on view in MOMA's 2017 show: *Is Fashion Modern?*

What makes a Lauren a Lauren? This is a paradox. If Ralph Lauren adapts past styles, and these change from season to season, how can there be a Lauren look? The answer is that Ralph Lauren is very clever: some costumes change, others remain constant.

The changes. A friend who owned an antique clothing store in Manhattan told me that Ralph's people would buy "period" pieces to match next season's "look." The pieces would be de-constructed so that their patterns could be replicated. The look is always correct—it is stylish in a very specific way. One season, an audience enamored of *Out of Africa* could also be costumed (expensively) in pale colors, cool linens, multi-pocketed vests and jackets. Another, romantics nostalgic for *Casablanca* could appear in hatted and gloved in chicly tailored, subtly structured dresses and suits with shoulders. The look I lusted for was Russian (pre-Revolutionary, Tsarist), with marvelously oversized white fur hats and muffs.

The constants. There is a reason why *Polo* is Ralph Lauren's trademark. *Polo* represents old-money. It is horsey (hand-made saddles and boots, jodhpurs and belts) and nautical (lots of blues and whites, vintage wooden yachts), very well made and very expensive.

Fashion II: Reconstruction

The same friend, Jana Starr (2017), who sold period pieces to Ralph Lauren explained reconstruction in a chapter on fashion in my first book. She said reconstruction is not interpretation—interpretation is updating antique styles using new materials. Reconstruction involves re-arranging actual antique materials in new ways.

What makes a Victorian wedding dress fashionable? Once Vera Wang introduced the strapless wedding gown, covered shoulders became *passe*. Since high necked, long sleeved, bustled Victorian wedding gowns were not selling, Starr began using their gorgeous hand-made lace in newly designed ones. There is a lot of lace in a Victorian wedding gown. Fortunately for reconstruction purposes, it was made in standard pieces to be assembled at the time of the wedding, and then taken apart and stored for another bride to wear. Jana respects the integrity of the intention and the material itself. She never cuts into it. Instead, she rearranges the skirt pieces (fan-shaped, longer in the back for the bustle) into simpler, sensuous shapes that drape from the bra line to the floor. Lace from the top pieces become sleeveless ones from which the drapery flows. Something old (the material). Something new (the interpretation).

There are two interesting asides here. One, Jana recently told me that the market for hand-made lace was shrinking because machine-made laces are now made better than they were. Since good lace is no longer scarce, lace gowns can be mass produced and sold at lower-priced chains like David's Bridal. Two, think of Kate Middleton's wedding gown—V necked, long sleeved. Look at the latest collections. Kate's look remains influential. Strapless is still stylish, but there are also styles with sleeves (shorter and longer) or little boleros, and also some that are sleeveless (like Jana's designs) but not strapless. I am waiting to see the variations on Meghan Merkel's much more sophisticated gown (designed by Givenchy).

What made 2017's fashion shows look like patchwork quilts? I graduated from Pratt, so I am occasionally invited to the fashion show featuring "collections" by about to graduate seniors in the fashion program. What struck me (aside from the extra sleeves and other gratuitous appendages) was the obvious patchwork construction of everything in the show. The head of the design program later told me that all of the fabric was antique, no new fabrics were used. (This is what makes it fit under reconstruction.)

I called Jana and asked where this all came from. Her answer was: take a walk on Madison Avenue and look in the boutiques. They are all showing un-wearable patchworks. It came from the necessity of being "fashionable." The problem is that it was neither beautiful nor wearable. It would certainly be noticeable if that was the wearer's goal. It will most likely not be back next spring.

Automotive, Etc

Another product category where new paths are "required and desired" is automotive. Although I (Pat) can't tell the difference between my car (a white Honda Accord) and all other small-ish (compared to SUVs) white cars in the parking lot behind ShopRite, I'm sure that there are more differences than the designs of their logos or on their tires.[1]

Etc. is an important, catchall, category. It includes all products whose consumers have come to expect (require/desire) novelty. It includes the latest iPhone as well as the most wanted (for the time being) toy, the most revolutionary face cream. Built-in obsolescence makes innovation necessary. Permeability makes it possible.

References

Gibbert, M., Goldenberg, J., Horowitz, R., Levay, A., & Mazursky, D. (2006). Finding your innovation sweet spot. *Harvard Business Review*, 120–129.

Starr, J. (2017, May 7). *Personal communication*.

[1]Aside from styling, there are the more recent changes in power, which are *desired but not required*. This puts them back in Category I: transitions from Same Path/Same Goal to New Path/Same Goal. Is your car a crossover, using gas and battery? Is it all-electric? As innovative as these changes may be, the overall goal (getting from place A to place B) remains the same, only the path changes. Same kind of thing could be said for the self-driving car—instead of car and driver, we have car *as* driver (again getting you from place A to place B).

Chapter 6
Category III: Same Path/New Goal to New Path/New Goal

Particular people will discover particular opportunities [if they are in] possession of the prior information necessary to recognize an opportunity...
Shane and Venkataram

There must be a new path for there to be an innovation.
Pat

Starting on an old path can lead you to a a new one.
Michael

The "prior information" referred to in the first quotation often takes the form of an existing (i.e., same) solution path that the innovator (or innovative team) notices as a promising way to start structuring a new one. *Start* is the critical word. Think back to Post-It Notes starting with a not-so-sticky glue that someone at 3M noticed as the start of a new product category.

In short, this is (like Category I) a ***transition*** category. An existing path suggests substitutions for forming a new (i.e., innovative) solution path. Or, said another way (as Michael did earlier), old paths became your point of departure.

Appropriately, since transitions are Michael's idea, most of the examples are also his. In the first, an existing product is modified; in the second, an existing procedure is modified; in the third, an existing product is repositioned; and in the fourth, existing products are combined to create an instant success.

Each example is labeled, appropriately we think, a point of department.

Point of Departure I: Modifying an Existing Product

Michael already described this transition, the metamorphosis of the Pressman into the Walkman (see Chap. 2). To remind our readers, Michael, I've copied the problem space from Chap. 1 (Table 6.1).

P. D. Stokes and M. Gibbert, *Using Paired Constraints to Solve The Innovation Problem*, https://doi.org/10.1007/978-3-030-25771-2_6

Table 6.1 Pressman to Walkman

Initial state: Pressman—portable recording tool			
Search space	**Constraint pairs**		
	Preclude	→	**Promote**
	Record and play back		Only play back
	Specialized tapes		Standard tapes
	Speakers		Ear phones
	Public listening		Private listening
Goal state: Walkman—portable, private listening tool			

As we all know, the Walkman morphed into devices that stream or download music. The devices have the same goal—a portable, private listening tool.

Point of Departure II: Modifying an Existing Procedure

Michael introduced this problem in Chap. 2, but made us wait for the solution. Here it is (adapted from Gibbert & Scranton, 2009).

In this example, the innovation problem involved jet propulsion technology. The blank check was the virtually unlimited (unconstrained) funding given the American team. In contrast, its German competitor was constrained by severely limited funding.

Let's start with the specifics of the problem. The big challenge in jet engines was and is to solve a performance dilemma: the more powerful the engine, the hotter it gets; the hotter it gets, the more frequent are failures due to material fatigue. At the end of World War II, several American teams under General Electric and several German teams under BMW and Heinkel were competing in a race to resolve this dilemma. The stakes were high, all parties recognized jet propulsion as the next "winning weapon" in the skies. The American team had a virtual blank check for buying whatever costly raw materials it needed to create the most heat-resistant alloys (indeed, the Cold War jet-propulsion development program cost the U.S. government nearly twice as much as the Manhattan Project). By contrast, the German engineers were denied access to state-of-the-art heat resistant alloys due to funding problems and postwar disruptions of international trade. They simply could not procure the required materials and had to make do with whatever was available.

This is a good story, Michael, we need more details.

Sure. In 1942, Herman Oestrich and his Munich team of BMW engineers and technicians faced a key problem in designing and fabricating a jet engine for the Luftwaffe: because of the war's sanctions, BMW was walled off from a variety of metal alloying elements and compounds, available only in Africa, Asia or the Americas. The resource constraints created Oestrich's team was working under created what seemed an impossible situation for fabricating aircraft engines. Denied access to key elements for alloys, due to war-related commercial blockages, and with the basic and low-alloy steels available, the plant could either produce engines that would be underpowered and hence deliver little or no substantive combat advantage (as they could not get hot enough to generate adequate thrust) or engines with adequate thrust, but prone to sudden and catastrophic failure (through component melt-downs).

Overcoming these constraints could profoundly affect the development of jet engines. Oestrich and his team reasoned that if there were ways to cool elements of the engine such that higher temperatures in working parts could be offset by cold airflows, the obstacle posed by unavailable heat-resistant alloy elements could be sidestepped. Further, if simple ways to generate this cooling could be devised, Goering's time pressures might be evaded, or at least, mitigated. After a series of trials, two technical maneuvers emerged. The BMW team split the incoming cold airflow to the compressor, one portion going around the outside of the combustion chambers and turbine wheel (the "hot section") to provide cooling and thus higher efficiency, and the rest into combustion. Also, another diverter sent a small proportion of the incoming air (2%) through internal channels which delivered cooling flows into the blades of the turbine wheel, which had to suffer the hottest temperatures. Into the blades? Yes, once they redesigned the blades.

Turbine blades are like scoops, slightly twisted to capture a bit of thrust and create a fan-like spin for the wheel and the shaft that drives the compressor. At the time, components suppliers either cast blades in molds (from liquid metal) or machined them (from solid blanks). The German engineers, however, took a different approach. Using sheet metal, folded over to create an inside channel, these innovators designed a turbine blade with a passage through which cold input air could be delivered. As an additional benefit, the production process for folded blades was simpler than either metal casting or metal grinding, speeding output and erasing the temporal tension. These solutions to Nazi jet development problems came too late to alter the war's outcome. BMW built some 500–700 of its Model 003 engines, but none of them powered a combat aircraft before war's end.

Actually, I believe that particularly when you cannot chose your initial state, the initial state seems to consist only of obstacles, constraints, and implausible situations, the key to innovation is to take it all in with an eye on unashamedly looking at the predicaments you are faced with (no heat resistant alloys, funding restrictions, extreme time pressure at the war's end). "Accepting" these initial conditions may hold the key to reassembling them in new ways to create a novel choreography for problem solving. Basically, the Germans bricolaged around the problem of melt-downs due to high temperatures not by creating more heat resistant alloys (for which there was no money) but by inventing more efficient ways of cooling the poorer alloys. Thus, the resource constraints characterizing the initial state provide the impetus for bricolage. If we cannot make heat resistant engines, we need to cool them better. In retrospect, it seems that simple!

Table 6.2 By-pass problem

Initial state: Material fatigue caused by heat			
Search space	**Constraint pairs**		
	Preclude	→	**Promote**
	Altering the alloy		Cooling the existing alloy
	Solid rotor blades		Hollowed-out rotor blades
	Metal casting		Metal folding
	Airflow only around blades		Airflow through blades
Goal state: By-passing material fatigue			

Michael, again let me try to turn your example into a problem space. The initial state for both American and German teams was heat-induced material fatigue. The German team had neither the funds nor the materials to develop a heat-resistant allow. Their goal (I'm using your term) became "by-passing" material fatigue. This precluded altering the alloy and promoted cooling the existing alloy. To do this, they precluded solid parts and promoted hollowed-out parts. To make the hollowed-out parts they precluded metal casting and promoted metal folding. This in turn precluded air flow only around the blades, and promoted air flow through the blades, which cooled them.

So, the problem space would look like Table 6.2.

One more thing, Michael. You mentioned that the bypass was an analogy of the cooling system in piston engines. This is a good example of "*what is needed*." I imagine that both American and German engineers were aware of the cooling system, only the Germans needed and so noticed the cooling system.

Yes, they both must have been aware of it, and yes, only the Germans needed and noticed it. "If we cannot get heat-resistant engines, we need to cool them better." In retrospect it seems that simple.

Sometimes it can be that simple. It all depends on what's in your toolbox.

And how you use your tools.

Point of Departure III: Repositioning an Existing Product

Clearblue is a company producing pregnancy tests. What are pregnancy tests good for, semantically and procedurally? The company started out marketing their products primarily to those that were hoping NOT to be pregnant. They soon realized, however, that another market segment is those that actually hope to BE pregnant. This involved changing the procedural memory of the product's usage (from hoping to see that gap to hoping to actually see the line connecting). Semantically, it suggests an important departure from a marketing communication perspective:

Clearblue became the centerpiece for a community of couples who actively want to have children, for connecting with others about the nine months of joy of expecting a child. It also changed the semantic field of products associated with Clearblue (from contraceptives to baby clothes, furnishing kids' rooms, and teddy bears). By creating a community of soon-to-be parents and those that just had children, it filled an important gap that a company operating after the first nine months (Pampers, say) would find difficult to address. A nice side effect might be making those that were hoping not to be pregnant a little bit more comfortable, understood, and even prepared for what expects them.

Michael, I love this example because changing the product's promise to a different audience was the innovation. That shift is what created the new path.

Correct. Clearview started out making pregnancy tests for those hoping they were not pregnant. Changing both the path (marketing to those who wanted to become pregnant) and the goal (wanting to be pregnant) made it possible for the company to move out of the contraceptives segment and into the expecting mothers segment.

Point of Departure IV: Combining Existing Products

This very short example is Pat's, who remembers succulent pork chops, potatoes, and sauerkraut made in her mom's pressure cooker. She doesn't have a pressure cooker, but she may get one of these Instant Pots.

The points of departure were the existing products that an Instant Pot combines: pressure cooker, slow cooker, rice cooker, and yogurt maker. As the problem space (Table 6.3) shows, the substitutions offer several advantages from space saving (single versus multiple appliances) to safety and time-saving. The Instant Pot has sensors that keep it from overheating or exploding (the exploding is why I never bought that pressure cooker). Melissa Clark, a food writer for the New York Times, reported cooking a pork shoulder in 90 min versus 7 h in a slow cooker.

Table 6.3 Instant Pot

Initial state: Multiple appliances			
Search space	**Constraint pairs**		
	Preclude	→	**Promote**
	Separate appliances		Single appliance
	Overheating		Sensors regulating temperature
	Exploding		Sensors regulating pressure
	Long cooking times		Shortened cooking times
Goal state: Single appliance			

In this case, several old paths became the point of departure. The new solution path (the innovation) combined existing (the cookers) and novel technology (the sensors) to produce a truly Instant success.[1]

Reference

Gibbert, M., & Scranton, P. (2009). Constraints as sources of radical innovation? Insights from Jet Propulsion development. *Management & Organizational History, 4*(4), 01–15.

[1]Amazon reported selling 215,000 Instant Pots on one day in 2016.

Chapter 7
Category IV: New Path/New Goal

To embark on new paths, they need to recognize that they are in the right place at the right time.
Ahuja & Katila

Boundaries focus attention.
Michael

Opportunity is the fairy godmother of invention.
Pat

This chapter focuses on two catalysts, opportunity and aspiration, that do not involve transitions. In cases of opportunity, something is *noticed*.[1] Generativity is a subset of opportunity. Here, the initial noticing leads to further, more highly focused noticing, which generates sub-goals, often in succession. Monet noticed the color wheel and painted light breaking up (in succession) on things, between things, and finally, sans things, by itself.

In aspiration, something is *next*. Innovators here aim to expand their domains, pushing the *boundaries* of what belongs. The painters who invented Pop Art *aimed* to replace Abstract Expressionism, the then-dominant style. Abstract Expressionism became the initial state in their innovation problem. It provided their preclude column, without which no Pop Art.

Since this is our closing constraint category, we emphasize the two things that we began the book with—paired constraints and boundaries.

Opportunity: Something is Noticed

Michael, I thought I'd start with a short example, putting the origins of a well-known success story into a single problem space. The success story is Starbucks. After Starbucks, I'll continue with *Only the NUMBERS Count*, the math

[1]As famously noted by Pasteur, noticing depends on accident (chance) and knowledge (the prepared mind).

P. D. Stokes and M. Gibbert, *Using Paired Constraints to Solve The Innovation Problem*, https://doi.org/10.1007/978-3-030-25771-2_7

program I invented using the constraint model, and perhaps you will tell us more about sous-vide…

Opportunity I: Starbucks

What happens when a coffee distributor walks into a Milanese espresso bar? If the coffee distributor is Howard Schultz, the answer is Starbucks (Table 7.1).

What Schultz *noticed* was an *opportunity* to turn a coffee bean chain into a coffee house chain. In the problem space for Starbucks, the two "chains" are the initial and goal states. A sub-goal was to make Starbucks a "third place" where customers regularly gathered (the other two "places" were work and home). The constraint pairs in the search space specify the kind of coffee house. Since I suspect that Schultz had at least one other source, the diner where Americans went for a cup of coffee, I included it in the search space. In a sense, both sources—the espresso bar and the diner—functioned as boundaries that structured the problem space and its solution path, each providing different opportunities for substitution.

Opportunity 2: Introducing Only the NUMBERS Count© in Kindergarten and First Grade

The innovation here is an early math program designed by a non-mathematician, me. What I *noticed* was an *opportunity* to turn early math programs that had more to do with words than numbers into a program that had everything to do with numbers, symbols, and patterns. The name of the program—*Only the NUMBERS Count©*—summarizes the goal: to make children fluent in numbers, symbols, and patterns (Stokes, 2014a, b, 2016). There were three things I had "at hand" to help me: the Japanese count I used when I worked for J. Walter Thompson in Tokyo, the design skills I acquired at Pratt, and, from psychology, my expertise in expertise.

Table 7.2 shows the initial to-be-precluded state as current curricula. Fluency was the goal.

The first thing at hand was the explicit base-10 count used by Chinese, Japanese, and Korean children. I think of it as the Japanese count because that is where I learned it. Table 7.3 shows the count (in English) from 10 to 30, which is called three-ten. Notice that only ten number names (1 thru 10) are used, and ten appears in every number above 10, emphasizing the base-10 iterations (10-2, 2-10-2, 3-10-2).[2]

---

[2]Importantly, the count solves what is called the place-value problem. If a child calls a number "two-ten-three," she will think of it as two 10s and 3 ones (the way Asian children do) and not as 31 ones (the way American children do).

Table 7.1 Problem space for Starbucks

Initial state: Coffee bean chain			
Search space	**Constraint pairs**		
	Preclude	→	**Promote**
	Selling only beans		Selling beans and brewed coffee
	Italian espresso bar		
	China cups		Paper cups
	Standing at bar		Seating
	Limited time		Extended time
	American diner		
	Coffees: limited choice (regular, decaf)		Extended choice (espresso, latte, etc.)
	Server: waitress		Barista
	Food: full meals		Limited menu
	Seating: Booths		Small tables, comfy chairs
	Counter by server		Counter by window
	Jukebox		Wi-fi
Goal state: Coffee house chain			

Table 7.2 Problem space for kindergarten and first grade

Initial state: Current curricula for kindergarten and first grade			
Search space	**Constraint pairs**		
	Preclude	→	**Promote**
	Non-numeric		Numeric-symbolic patterns
	English language count		Explicit base-10 count
	Multiple manipulatives		Single manipulative
	Split practice		Deliberate practice
Goal state	New curriculum (*Only The NUMBERS Count*©)		
Criterion	Fluency: thinking in numbers, symbols, and patterns		

The second was the design skills in my tool box. Figure 7.1 shows the manipulative—called the Count and Combine Chart—for numbers 1 through 10.

Figure 7.2 shows part of the chart for the numbers 10–13 (ten-three). Notice that the tens are represented by single blocks marked "10." Notice how the similarity in the charts helps make the iterative patterning of the count explicit.

The third thing was expertise in expertise. What I knew was that experts solve problems using meaningful patterns in their domains of expertise. For mathematicians, the patterns involve numbers and symbols and relationships. "Could children, practicing primarily with numbers and symbols, learn to think like

Table 7.3 Explicit base-10 count from 10 to 39 (three-ten-nine)

Tens	Twenties	Thirties
10 ***ten***	20 two-***ten***	30 three-***ten***
11 ***ten***-one	21 two-***ten***-one	31 three-***ten***-one
12 ***ten***-two	22 two-***ten***-two	32 three-***ten***-two
13 ***ten***-three	23 two-***ten***-three	33 three-***ten***-three
14 ***ten***-four	24 two-***ten***-four	34 three-***ten***-four
15 ***ten***-five	25 two-***ten***-five	35 three-***ten***-five
16 ***ten***-six	26 two-***ten***-six	36 three-***ten***-six
17 ***ten***-seven	27 two-***ten***-seven	37 three-***ten***-seven
18 ***ten***-eight	28 two-***ten***-eight	38 three-***ten***-eight
19 ***ten***-nine	29 two-***ten***-nine	39 three-***ten***-nine

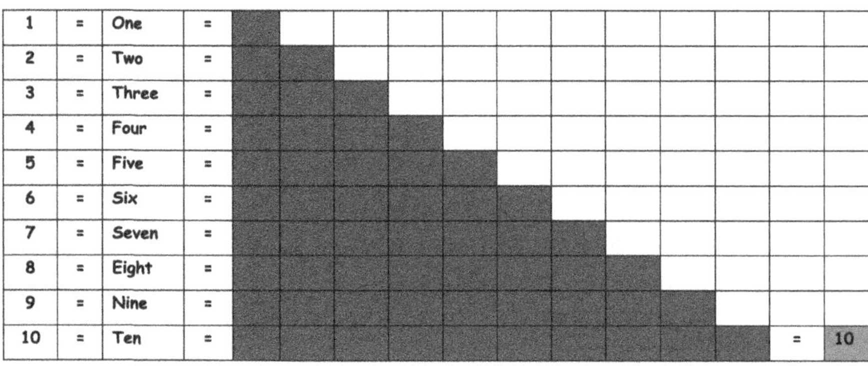

Fig. 7.1 Count and Combine chart 1–10

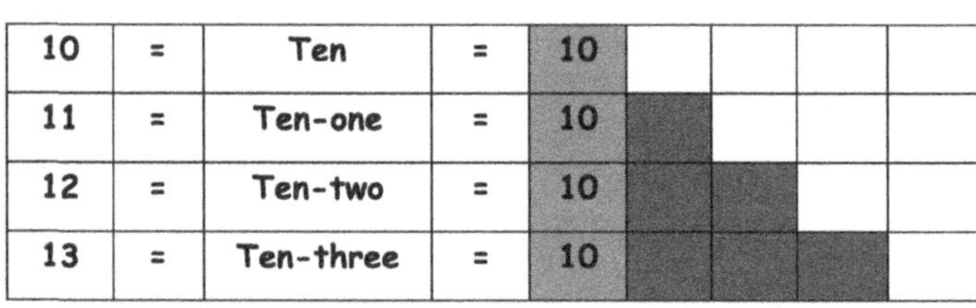

Fig. 7.2 Count and Combine chart 10–13

mathematicians?" was the challenge. I also knew about deliberate practice which is highly focused (on specific aspects of a skill), continuous, incremental (developed in successive steps) and highly varied. For example, children practiced the pattern of the base-10 count by chanting and reconstructing the chart. The chart went this way: *Number one same as word one equals one block, Number two same as word two equals two blocks....* The children practiced the structure of base-10 solutions for addition by combining and recombining the blocks (for numbers and symbols) to make different "combinations."[3]

Pat, did children learn to think like mathematicians?

Gratifyingly so Michael. Kindergarteners and first graders exceeded Common Core standards for their grades, meeting second grade standards for Operations and Algebraic Thinking, and Numbers and Operations in Base Ten.

You know Michael, re-reading this section, I realized that what I want our readers to remember are not the specifics of the new math program, but the importance of the old one. Without current math curricula to work against, I would have had no preclude column, no place to start the substitution process. The take away is two-fold: *every new idea starts with an old one*; the initial state, a boundary in its most basic sense, is critical.[4]

So what is the initial state for sous-vide as a cooking (and not re-heating) technique?

The traditional technique for making foies gras.

Opportunity 3: Sous-Vide II

Sous-vide ***cooking*** *in its literal sense and modern form plastic vacuum packs is said to have been developed by Georges Pralus, a French chef, who noticed an opportunity to adapt sous-vide (a re-heating technique) to solve his foies gras problem.*

I love it. We should all have foies gras problems!

The story is that Chef Pralus developed the new sous-vide in 1974 while working at the Restaurant Troisgros in Roanne, France. Foies gras is the result of torturing geese (traditionally) or ducks (the cheaper variant) by force-feeding them maize until their livers become fatty and up to ten times the size of a normal liver. Why ducks and geese? These animals travel enormous distances when migrating and Mother Nature has developed in them an incredible capacity to store extra nutrients in the form of fat in the liver. The liver of course always stores nutrients (mostly glycogen) for later release. What makes the livers of geese so exceptional is their capacity to swell up more than the liver of any other animal without the goose dying (the liver, in theory, can go back to normal again but alas most of these geese do not see this happen).

---

[3]The word combination was used to reinforce the idea that numbers are combinations of other numbers. There are, for example, sixteen different addition combinations for 5, and thirty-two for 6.

[4]I also realized that this may be more than enough math for one chapter. So, ***Solution by Substitution II; Second and third grades*** (which is far more complex) is being moved to the back of the book (See Appendix A).

The challenge—that is the problem—with cooking foies gras is that it literally melts (a bit like butter), precisely since it is mostly fat, held together by a bit of liver tissue. The trick in preparing a good terrine de foies gras is to break up the liver in smaller (walnut-sized) pieces, marinate them (e.g., in a reduction of Port and black truffles) and then stuff them in a terrine or other dish, put the dish in 60° (Celsius) water and let the whole thing amalgamate very, very slowly. The result, when you cut a slice off, is a beautifully marbled (each walnut-sized piece of liver has a purplish-black ring from the marinade and ideally is rosy pink inside) culinary delight in its purest form (although best avoided for animal welfare, if you ask me). The sous-vide approach works in exactly the same way but avoids the frequent disappointments which come from the liver shedding too much fat. This happened to me once, when I left the foies gras for slightly too long and ended up with a (very expensive) mass of yellow fat floating on grey shreds of liver. The sous-vide method here was an ingenious (and very cheap) new way of making sure that this exclusive raw material could be exposed directly (only the plastic membrane separating the liver from the heat) to much lower but not too-low temperatures (keeping water temperatures too low results in an undercooked, blood-laced product).

After having been tested for foies gras, sous-vide cooking appeared in kitchens across the globe for almost anything from meat to vegetables.

Three questions. One, how do you test the water temperature? Two, what are the boundaries for any cooking technique? Three, do you have a recipe we (our readers and me) could try at home?

Answer number one: if water temperature is critical (as it is for foies gras), the best technique is simply to use a thermometer. Mind you, you still need to be there in person check in once in a while or you wind up with the dreary result I described elsewhere.

Number two: the boundaries for any cooking technique (and coincidentally, for any innovation, a.k.a. new recipe) are essentially only three. For one, it is the availability (or lack) of ingredients. Second, it is quite simply time. Consider ravioli, and in particular, those filled with meat. These were born of the necessity to feed large families. Meat was scarce and precious. So the original ravioli were filled with leftover bits from the last Sunday roast or roasts (in professional kitchens, there tend to be leftovers from several roasts). A great illustration are the 'agnolotti del plin ai tre brasmati.' These are a local specialty here in Piedmont and are truly delicious. Basically, they are very small ravioli filled with bits and pieces from whatever roast/stew you have in the corner of your fridge. These are cut up/minced finely, mixed with egg yolk and parmesan and then used as a filling.

YouTube has some misguided videos about making agnolotti. The problem with their recipes is that they are not getting the crucial idea right: the agnolotti are a method for avoiding food waste. That is, the idea is NOT to prepare three different roasts (each using minute amounts of meat) and then chopping them up. Can you imagine how long it takes to prepare three separate roasts before even thinking of filling the agnolotti? It completely misses the second amendment for cooking innovation: use necessity, avoid waste—and accept the fact that it will take time (though less time than when you start roasting from scratch).

Three: the recipe. I strongly suggest that you (and our readers) try this delicious method of using leftovers. Simply stuff them (in tiny heaps) on a sheet of fresh pasta (one small egg per person, plus enough durum wheat flour to make an elastic dough). You then distribute the little piles of filling in regular intervals (not too close together), wet the pasta around the filling, and put another layer on top—then you simply cut out the individual ravioli. You can cut them square or use a small wine glass to cut out round ones. This leaves cut-off bits, which—in the Italian tradition—you do NOT throw away, but keep and cook separately as maltagliati (literally, 'badly-cut pasta', or simply cut-offs).

You can even prepare them in advance and freeze (individually, on a flat tray to keep them from sticking together). Fresh or frozen, just boil for a few minutes and toss into a pan with some butter, herbs, garlic, or any other sauce you may prefer. Buon appetito.

One last question: is the sheet one big piece? And when does it get cut up into separate ravioli?

Yes, it is one big sheet which comes out of your pasta machine (which conveniently produces nice long, rectangular sheets). If you (or our readers) are unfamiliar with the technique, I suggest typing "pasta fresca" into YouTube to get some guidance from the real Italian mammas—they will be in Italian, but the visuals are all you need. And they beat the Jamie Oliver stuff which has become the mainstay with pasta-fresca newbies this side of the Atlantic Ocean.

*Opportunity 4: Thumb*roll

Thumb*roll* is an app invented by a former research assistant[5] in my lab who began with an idea (showing how something is done will engage someone in what they are trying to do) and a prototype (a platform that allowed viewers to thumb through a series of images as slow or fast as they choose), but no audience until she *noticed* an *opportunity*—helping healthcare trainees and professionals master basic (and latest, best) procedures faster and more efficiently.

A completely specified problem space for thumb*roll* was also created by Ariella. Table 7.4 shows my abbreviation. The initial state was existing videos of medical procedures. The preclude column characterizes those videos. The promote column shows the substitutions that let users preview, review, and—something impossible with standard videos—practice the procedures.

[5]Ariella Salimpour.

Table 7.4 Problem space for thumb*roll*

Initial state: Teaching videos			
Search space	**Constraint pairs**		
Categories	**Preclude**	→	**Promote**
Media type	Video		Series of images, brief annotations
Learning type	Passive		Active engagement
Viewing time	Set		View at your own pace
Content	Excessive information		Optimal, single focal point
Time	Time consuming		Quick
Senses	Visual and auditory		Visual and tactile
Viewer's position	Stationary		Stationary or mobile (review on-the-go)
Goal state: Faster, more efficient way to master medical procedures			

If you're a medical professional, your procedures must be perfect. A procedure is an action pattern—suturing a wound, inserting an IV, riding a bicycle—that, once acquired, runs off automatically. What actually runs off is a motor program stored in the basal ganglia or cerebellum. The reason why practice should be perfect is this. The brain stores everything you practiced, including your early, inexact steps. Result? The inexact steps show up now and then.

Thumb*roll* helps with this problem in two ways. One, it provides correct, step-by-step practice. Two, it provides correct, step-by-step, review. To see how thumb*roll* works (and why it's in over 100 medical schools and residency programs in the USA), just visit their web site (http://thumbroll.com).

Aspiration: Something is Next

Two of my examples (no surprise) involve art: two fine (Pop Art and one Pop painter, Roy Lichtenstein), the other commercial (Pixar). The third is an innovative way to teach history. Michael will tell us about the BlaBlaCar.

Each of these innovators aimed at creating the next "big thing." They were all successful.

Aspiration 1: Pop Art

How did Pop Art start? The story goes something like this.

A group of young painters, realizing that fame and fortune required <u>not</u> being second-generation Abstract Expressionists, make a list of the things that

Table 7.5 General problem space for Pop Art

Initial state: Abstract expressionism (emotional style)			
Search space	**Constraint pairs**		
	Preclude	→	**Promote**
	Abstraction		Realism
	Emotion		Detachment
	Improvization		Pre-planning
	Complicated form		Simple, hard-edged shapes
	Virtuoso brushstrokes		Flat, primary colors
	Fine art imagery		Commercial imagery
Goal state: Pop Art (ironic style)			

characterized Abstract Expressionism?[6] Notice how Abstract Expressionism acts as a boundary—all the substitutions stem from its defining characteristics. They then made a list of opposites (as substitutes). Table 7.5 shows the constraint pairs that characterized this shift from the emotionally "hot" to the ironically "cool."

A second question: if all the Pop painters subscribed to the same (slick, simple, cool) set of constraints, how did the work of each become recognizably theirs? By incorporating borrowings: all from commercial art, each unique to a specific painter's tool box. For example, James Rosenquist had been a sign painter, his paintings looked like billboards; Andy Warhol had been an advertising illustrator, his paintings replicated packaging.

The Pop painter whose career I will expand on about is Roy Lichtenstein

Aspiration 2: Roy Lichtenstein

Lichtenstein's recognizable comic book imagery began with drawings of Mickey Mouse made for his children. I trace his innovative development in three stylistic phases. Images from each are easily found on line.

Solution by Substitution One: Making comics monumental. "Monumental" comes from the artist himself: "I take a cliché and organize its forms to make it monumental." It is also what I think of as his sub-goal in this phase. The italics in Table 7.6 show the changes from the general Pop problem space to phase one for Lichtenstein.

Pat, the last substitution is Ben-Day dots. What are Ben-Day dots?

I should explain that. The dots are Lichtenstein's signature material. Ben-Day dots were widely used in the 1950s and 1960s for printing comic books. They come

[6]In a recorded interview, Roy Lichtenstein told John Coplans that "desperation" spurred the shift: "There were no spaces left between Milton Resnick and Mike Goldberg…" Resnick and Goldberg were two second-generation Abstract Expressionists. Resnick was one of my teachers at Pratt.

Table 7.6 Problem space for phase one

Initial state: Abstract expressionism (emotional style)			
Search space	**Constraint pairs**		
	Preclude	→	**Promote**
	Abstraction		Realism
	Emotion		Detachment
	Improvization		Pre-planning
	Complicated form		Simple, hard-edged shapes
	Virtuoso brushstrokes		Flat, primary colors
	Fine art imagery		*Comic book imagery*
	Fine art materials		*Ben-Day dots*
Goal state: Pop Art (ironic style)			
Sub-goal: *Making comics monumental*			

on sheets, each covered with dots of a single size, evenly spread out across the sheet. The dots are transferred to a different surface by rubbing the back of the sheet. Lichtenstein's comic style paintings were very large, the dots he used were correspondingly (and recognizable) very large (Table 7.6).

While Lichtenstein started with Mickey and Donald, characters from children's comics were soon replaced by action and romantic images. *As I Opened Fire* (1964) borrows from low-brow art with its blazing weapons, explanatory text ("The enemy would have been warned") and exclamatory sounds (Brat!). It also borrows from high-brow art with its triptych format. Lichtenstein's *Drowning Girl* (1963) is immersed in what may be her own tears, certainly in her own over-romanticized thoughts: the text balloon reads "I'd rather sink—than call Brad for help!".

Solution by Substitution Two: Re-seeing art history. In this second, extended stylistic phase, the artist re-viewed and re-imaged art history, the boundary that all new styles aim to expand. For viewers to realize what he was doing (adapting the style), they had to recognize his sources. Thus, content was copied. *Haystack* (1969) is a black and white simplification of middle-period Monet. *White Brushstroke I* (1965) turned Abstract Expressionism's improvization into iconography. The brushstroke (huge and dynamic, reminiscent of De Kooning) is dramatically outlined in black and filled with a flat, single hue (white); its background is a field of black-on-white Ben-Day dots.

Table 7.7 shows the problem space for phase two.

My attempt to copy *White Brushstroke I* is shown in Fig. 7.3.

Solution by Substitution Three: Looking like the thing itself. The primary "things" in this phase were mirrors and entablatures. Again, the constraint pairs (shown in Table 7.8) shifted with the sub-goal.

The mirrors and entablatures became, like Monet's poplars and haystacks, subject to serial treatment. Lichtenstein painted 40 mirror images. The titles name their subject (e.g., Mirror No. 1). The canvases take on mirror shapes: ovals, rectangles, circles, single or multiple panels. However, unlike real mirrors, there are

Table 7.7 Problem space for phase two

Initial state: Making comics monumental			
Search space	**Constraint pairs**		
	Preclude	→	**Promote**
	Text		Pure imagery
	Changing the content		Copying the content
	Adopting the style		Adapting the style
Goal state: Ironic style			
Sub-goal: Re-seeing art history			

Fig. 7.3 Pat's attempt to copy White Brushstroke I

Table 7.8 Problem space for phase three

Initial state: Re-seeing art history			
Search space	**Constraint pairs**		
	Preclude	→	**Promote**
	Content		Form
	Ornate		Pristine
	Color-full		Limited palette
	Changing motif		Series of same motif
Goal state: Ironic style			
Sub-goal: Looking like the thing itself			

no images reflected in the black and white Ben-Day dot patterns. This too reminds me of Monet. Was Lichtenstein painting light, not on the reflecting surface of a pond, but on the reflecting surface of a mirror?

The entablatures are also empty, but less enigmatic: simplified, stylized images of conventional architectural details. An entablature is a three-part horizontal band supported by columns. The Greeks filled the middle row (the frieze) with sculptures; Lichtenstein's precluded figuration in the paintings. Like the mirrors, the shapes of the canvases take on the horizontal shape of their subjects. Each painting (in either series) takes the artist at his own word: "It doesn't look like a painting of something, it looks like the thing itself."

Look for the mirrors and entablatures on line. Look closely. These are serious paintings that—surprisingly—still satisfy the constraint pairs that started Pop.

Aspiration 3: The BlaBla Car

Michael, from what little you've told me (so far) about the BlaBla Car, it sounds as if its inventors aspired to put car-and-driver into a whole new category. It's time to tell us more.

Yes, exactly. To start off, the BlaBla Car driver operates in a shared economy with a two-sided market. That is, someone shares their car for a certain ride. The driver is part of a community, and as such may also choose to be a passenger. Hence the two-sided nature of the service. The main thing, however, is that the service is not 'just' about transportation. It is also about meeting new people.

The social dimension is what separates BlaBla from say, Uber, which is not about the blabla (i.e., the communication), but simply about the transportation. Depending on how talkative you are (and want your passenger to be), you can register as a blabla person, or indeed a blablabla individual. Now, the social dimension is not to be confused with a dating service. BlaBlaCar is not Tinder, so women who want the blabla (or even the blablabla), but not a dinner invitation after the ride can choose the 'Rose' (women only) version.

Is there a fee for joining the community? Do people share tolls and gas costs? Or do you take a passenger only when they are going where you are going?

There is no fee for joining the community. The driver asks for a 'fee', which typically is the rider's share of the cost of gasoline and other expenses. And of course, you take only passengers when they are going where you are going. Basically, you advertise your itinerary on the platform, quote the fee, and see who wants to tootle along. The platform charges a percentage of the overall fee, and allows you to 'rate' your passenger and your driver.

Aspiration 4: Pixar

Pixar the winner of *The Economist's* 2014 prize for corporative innovation is a company, a collaboration, not an individual. Its much admired, acknowledged source (and primary boundary) was Walt Disney whose animation studio provided both

Table 7.9 Problem space for Pixar

Initial state: Walt Disney style animation			
Search space	**Constraint pairs**		
	Preclude	→	**Promote**
	Illustrator-generated animation		Computer-generated animation
	Characters acting like humans		Characters with human-like intentions/emotions
	Magical "saves"		Problem-solving "saves"
Goal state: Computer-generated animation			
Sub-goal: ... of emotionally compelling stories			

what Pixar promoted/worked with and what it precluded/worked against (Capodagli & Jackson, 2010; Catmull & Wallace, 2014). What it promoted was Walt's way of working. Ed Catmull, president of both Pixar Animation and Disney Animation, put it this way: "Disney's animators were at the forefront of applied technology: instead of using existing methods, they were inventing ones of their own... Every time some technological breakthrough occurred, Walt Disney incorporated it and then talked about it... in a way the *highlighted the relationship between technology and art*" (p. 8). The italics are mine. They highlight the catalyst and the core of Pixar's development which, ironically, required replacing Disney's technology and the sorts of stories that technology animated. The new goal (computer-generated animated) came with a sub-goal: emotionally compelling stories.

Pixar's dual innovation is condensed into the single substitution matrix shown in Table 7.9.

The first constraint pair condenses the multiplicity of techniques that were precluded and promoted. Traditional techniques included: illustration, stop motion, blue-screen matting, multi-plane cameras, and xerography. Developed to improve film editing at Lucasarts, the original Pixar Image Computer scanned film strips, added special effects and scanned the combination back in digital form. While *Toy Story* began, as all Pixar-animated films do, with hand-drawn storyboards, its curved shaped, realistically shaded imagery was generated entirely by the computer.

The second replaces characters like Mickey and Goofy—who dress, and walk, and talk like humans—with characters who experience issues that are emotionally compelling because they are ones that we, as humans, also face. Take Marlin (a fish), an over-protective parent who tries to protect Nemo, his only son, from the dangers of the ocean, or Bob Parr (a super-hero) living a boring "normal" life, who jumps at the chance to be Mr. Incredible again.

Other well-loved Disney characters—Cinderella, Snow White, Beauty—came from fairy tales. Since the tales were magical, so were the "saves"—a fairy godmother, a good fairy, a magic mirror. The third constraint pair substitutes problem-solving saves—helpful conversations, insights and intuitions, all based on experience—for magical ones. Take Dory talking to Marlin, who is himself afraid of the ocean. Listen to what Dory says: "Well, you can't never let anything happen to him. Then nothing would ever happen to him. Not much fun for little Harpo" (*Finding Nemo*, 2003). This is profoundly practical in suggesting a different parenting style. It is also exactly the kind of difficult, emotionally charged conversation you or I might have with an over-protective friend.[7]

*Pat, you're the psychologist. What did you think about **Inside Out**, the one about emotions?* I thought it was terrific, engaging and emotionally correct.

Aspiration 5: Re-enacting the Past

Reacting to the Past, an innovative way to teach history, was invented by a colleague at Columbia, Mark Carnes. Mark is a historian who wanted to make his subject intriguing and involving to undergraduates. His innovation was expanding the boundaries of the history classroom, retaining the history but having students re-enact (and not just read about) historic conflicts. As a psychologist, I know how difficult it is to understand how anyone is feeling or thinking, so how—I asked Mark—can a student act like a Calvinist at the time of Anne Hutchinson? Mark's answer was, teach a section and find out. I found out, and keep finding out. Here's a much abbreviated problem space (Table 7.10).

In the version that I teach, each student is given a role and a goal in three different conflicts that we re-enact. I see my role as teaching them how to find and focus on a single argument (this comes from advertising, of course) to advance their positions. In addition to speeches, my students also write concise, single-focus op-ed pieces for the Athens, Beijing, and Boston Times. Mark was right—you can put yourself in someone else's place *if* that place is clearly specified. You will also

Table 7.10 Reacting to the Past

Initial state: Traditional history class			
Search space	**Constraint pairs**		
Retain	**Preclude**	→	**Promote**
Primary texts	Passivity		Participation
	Lectures		Role-taking
	Neutrality		Position taking
	Papers		Speeches
Goal state: Interactive history class			

[7]I have to mention a different kind of friend here, a collaborator named Ben Xue. Ben is a Columbia College graduate on whose innovative independent project this section on Pixar is based. Thank you, Ben.

find it easier to argue and write if your position is clearly specified. Students and faculty (at many schools) learn much from re-enacting the past.

Well Pat, since your position (on the model) is clearly specified, did you find it easy to argue for/write about it for this book?

Relatively speaking, yes, but what made it fun was having something/someone to react/argue against. Remember you had to argue me through the expansion of the model.

Has anyone else done that?

Other collaborators and competitors, some borrowed from, some argued with, some aligned with. Some I'd like us to talk about in the next chapter.

Are the competitors other models?

Some are.

References

Capodagli, B., & Jackson, L. (2010). *Innovate the Pixar way*. New York: McGrawHill.

Catmull, E., & Wallace, A. (2014). *Creativity, Inc. Overcoming the forces that stand in the way of true inspiration*. New York: Random House.

Stokes, P. D. (2014a). How early is early enough? Solving the place-value problem in first grade. *The New Jersey Mathematics Teacher, 72,* 30–40.

Stokes, P. D. (2014b). Using a creativity model to solve the place-value problem in kindergarten. *International Journal of Creativity and Problem Solving, 24,* 101–122.

Stokes, P. D. (2016). Thinking in patterns to solve multiplication, division, and fraction problems in second grade. *Journal of Mathematics Education at Teachers College, 7,* 1–10.

Chapter 8
Evaluating Usefulness: Other Models and Ours

The question is: which strategy?
Pat

The question is: which detour?
Michael

Every innovation model is both a strategy and a detour from another strategy. In this chapter, the answer to our questions (which strategy? which detour?) is based on utility. Which *works* best is the real question. We consider the alternatives in alphabetical order, closing with the unique contribution of paired constraints to solving the innovation problem.

Other Models

Blending, Breaking, and Blending

We spend more time on this model than the others both because it is the newest and because it is so well written that the writing obscures problems with the argument. The model was introduced in a 2018 book, *The Runaway Species*. Its authors, Eagleman and Brandt (2017), label three tools the brain uses to "alter what it already knows." The tools are Bending (making over an existing prototype), Breaking (taking a thing apart and re-assembling its fragments), and Blending (combining multiple sources). The basis of all three is, obviously, *borrowing*, which makes an innovator's tool box bigger. There are two problems which limit the model's utility.

The first is that ***examples are not explanations.*** Consider Monet's series paintings. In the chapter on bending, they are explained simply as "representing a prototype in new ways" (p. 55). If the prototype were simply the subject (the façade

P. D. Stokes and M. Gibbert, *Using Paired Constraints to Solve The Innovation Problem*, https://doi.org/10.1007/978-3-030-25771-2_8

of the Rouen cathedral), then bending does describe using a different palette (color assortment) to vary the painting, but …

Wait, if the prototype isn't the first painting in the series, what is it?

It's the style, the product of Monet's second innovation when he (to use the book's categories) *broke*-up or *broke*-from the style of his first phase.

I thought we were talking about bending.

So did Eagleman and Brandt. Let's go back to Monet's first stylistic phase. The innovation we call Impressionism was the product of precluding specific elements of the then-dominant painting style (e.g., light-dark contrasts) and promoting substitutes (contrasting hues) to show how light broke up *on* things. The series paintings shown are the products of his second innovation, showing how light broke up *between* things.[1] This, I think, fits more appropriately under breaking, "the brain's technique of omitting some pieces and keeping others" (p. 86). What Monet kept was the idea of light breaking up, among the things he omitted/precluded were local color, the substitute for which was shared color and also separate brush strokes, the substitute for was a web of the shared colors.

So, the series paintings are examples of breaking, not bending.

Actually both, depending on your focus.

I see where you're going—is the second problem that ***the tools are interchangeable?***

I think so.

So, are they useful?

Far less than they are supposed to be. In fact, the author's (very general) suggestions for "exercising the creative mentality" (e.g., don't stop at you first solution, generate lots of ideas) have little to do with their tools.

Which are not tools…

But which provide lots of wonderful examples that could be *usefully operationalized* using the paired constraint model.

Boundaries

In Paul Cilliers' model (1998), boundaries are both the problem and the solution. Thanks to Michael, the constraint model *incorporated* the idea: boundaries *define* both *domains* and the *initial state* of an innovation problem. Chapter 7 pointed to multiple examples of this. In the Starbucks case, the espresso bar and the diner were boundaries that structured the problem space and the solution path by providing opportunities for substitution. In the case of Pop Art, Abstract Expressionism was the boundary without which Pop could not have developed. All the substitutions (the solution path) that defined Pop depended on precluding the defining characteristics, the boundaries, of its predecessor.

---

[1]Between things is the atmosphere or as Monet called it, the envelope (see Preface).

Bricolage

Bricolage is generally understood as a tool to surmount resource constraints. In paired constraint parlance, bricolage provides substitutes. One makes do with what is at hand. Michael has written about bricolage in the context of scarcity. Pat has written about bricolage in the context of Matisse's late cut-outs. Confined to bed, Matisse could not stand and paint with brushes at an easel. Making do with what he could do, Matisse cut shapes out of colored papers and had them pinned or pasted larger sheets, also of colored paper. Painting (plus brush, canvas, etc.) was precluded. Cutting-out (plus scissors, paper, etc.) was promoted. Bricolage is easily *incorporated* into the paired constraint model.

Collaboration

Collaboration, as amply demonstrated by Isaacson (2014), can be critically important to innovation. Collaborators *borrow* expertise from each other. Their borrowings, like bricolage, make each one's tool box bigger. A jazz ensemble (a metaphor for organizational innovation) depends on collaboration between musicians, each of whose contributions makes the group's tool box bigger.

What collaboration models focus on is group interactions. What the paired constraint model focuses on is the actions themselves, the preclude-promote process that structures the solution path. We think this makes the paired constraint model *more useful* for innovators.

Design Thinking

Design Thinking is promoted as a problem-solving model (Liedtka, King, & Bennett 2013). The problem is with the thinking behind the model. Design Thinking itself is not new. It's what done (and in far fewer steps) in advertising agencies and design firms like IDEO. It's learned, by doing, in a disciple. What this means is that Design Thinking is not as teachable as its promoters promise. In Chap. 4, I reported sampling a Design Thinking course on Coursera. Perhaps it would not have seemed so over-complicated and over-thought to someone who hadn't been educated at Pratt, JWT, and Ted Bates. Also, Design Thinking (as presented in the model) is, in many ways, amateurish. Whoever came up with "what WOWs" would not have gotten a job at any agency I worked for.

In short, Design Thinking (as presently promoted) is *not useful.*

Paths

Path-deepening and path-creating. In Ahuja and Katila's (2004) model, innovations are the product of two distinct paths. One resolves existing problems in new ways. They call this path-deepening search; it corresponds to *new path/same goal* in the paired constraint model. The other, which solves new problems, is called path-creating search. It corresponds to our *new path/new goal* category. In both cases, the innovation is (as always) the new path. In the first, it accomplishes an existing goal. In the second, it specifies a new one. What is missing from the model—and what the constraint model *more usefully* supplies—is the actual path-creating process.

The path that prevails. Garud and Karnow's (2002) description of an actual innovation process (in wind turbine technology) parallels the paired constraint model in two ways. First, it involves *solution-by-substitution*. The teams proceeded, they tell us, by successively reaching and precluding their own interim approximations. Both pairs and paths have parallels in their description of solution paths "*constraining*" (precluding) and "*enabling*" (promoting) team activity. Second, it shares our premise that the goal is reached when "one path comes to prevail over others." The path that prevails specifies the criterion: this is *how* a wind turbine works. In our model, that path is the completely specified promote column that transforms the initial into the goal state. Despite the parallels, the paired constraint model is *more useful* in that it provides a general template (not a single exemplar) for innovation.

Redesigning routines. Gupta, Hoopes, & Knott (2015) reconstructed a redesign process in which a baseline routine for a single location was transformed into a routine for a large-scale rollout. Several elements in the process parallel parts of the constraint model. First is "identify an existing routine blueprint." This is the initial state. Second is "decompose the routine into elements." The elements are the items in the preclude column. A third is "propose alternative means." These are the substitutions in the promote column that will specify the routine for the rollout (the goal state). As in the previous model, despite the parallels, the paired constraint model is *more useful* in that it provides a general template (not a single exemplar) for innovation.

Strategic Insight

This is the model most closely related to ours. Duggan (2007) also presents a concrete framework which he calls an insight matrix. The first step in filling in the matrix is identifying the problem. This is followed by breaking the problem into categories and then identifying precedents (things that have worked for other businesses) to fill in the categories and provide elements for combination. Precedents are *borrowings*, things that have solved similar problems for other innovators and businesses. Each precedent is partial, filling in one part of the matrix. The process involves selecting (*precluding/promoting*) and combining possible precedents. Duggan argues (as does Pat) that the goal is only defined when it is realized. In the paired constraint model, this happens when the promote column is fully specified; in the matrix model, when the matrix is fully specified.

The paired constraint model is simpler than the matrix model because it is more specific. You start with what you want to restore, refine, or replace. This becomes the initial state. The initial state identifies the elements to be retained or replaced. Because these are given, the search for substitutes is far less complicated. The promote column is the innovation, itself readily available to serve as the initial state for the next innovation. The simplicity, which makes the model accessible, can also make it *more useful* for innovators.

Pat, I know you like Duggan's model. Can you think of specific instances where ours would be more useful?

Sure. Instance one: there are *no precedents* at hand. Go back to the Monet example. The idea of breaking up light came from Chevreul's color wheel but what the color wheel provided was a problem (showing how light breaks up) not a precedent for its solution. Since the problem was new, there were no other precedents to be borrowed for its solution. Monet had to solve the problem by substitution, not by combination. So, when there are no precedents, the constraint model would definitely be more useful.

Instance two: *identifying precedents*. For example, Duggan listed two precedents for Starbucks: Seattle high-quality coffee-bean chain and Italian coffee bar. I added a third, the American diner. I also used the paired constraint model to specify what was retained, precluded and substituted for each precedent. For example—using the Italian coffee bar as precedent—in Starbucks seating replaced standing, paper cups replaced china ones, and extended time replaced limited time.[2]

Now Michael, since you're the expert on thematic thinking….

Thematic versus taxonomic.

Thematic (vs. Taxonomic) Thinking

Since most managers think in taxonomies, I'll start with taxonomies.

Taxonomies are mental models which sort objects into categories. The sorting is based on similarity. Simply put, the degree of similarity depends on the extent that two items are from the same superordinate category.[3] *The problem is that two*

[2]See Pat's problem space for Starbucks in Chap. 7.

[3]Perhaps due to the intuitive appeal of the idea that similarity increases with common, and decreases with distinctive, features, it has served as the basis for key frameworks like strategic relatedness, taxonomies such as the Standard Industry Classification system and patent classification systems such as the International Patent Classification. For example, IPC category F02 (combustion engines) contains internal-combustion piston engines, gas-turbine plants, jet-propulsion plants, and so on.

things can be similar in one way, yet dissimilar in another. For example, apples and oranges both belong to the superordinate category, fruit, and as such fairly similar. However, when asked how similar the two are for having a healthy snack while driving a car, participants rate them as very dissimilar.

The problem for mangers is this. It is easier to interpret an opportunity or threat that is "close" (i.e., in an industry with similar factor inputs, distribution channels, and demand structures) rather than "distant." However, managers have traditionally been admonished to search in categories that are distant rather than close. The problem here is that searching in distant areas makes the number of possible opportunities theoretically infinite. So where do you start? And even more importantly, where do you stop in this infinite space? The solution, I think, lies in precluding the idea of a one kind of similarity and promoting the idea of two.

Precluding, promoting, I like that. What are the two kinds?

Not surprising given the title of this section, one is taxonomic, apples and oranges are fruit. The other is thematic, apples and oranges function differently in a specific scenario (eating while driving). Apples and oranges are taxonomically similar, but thematically dissimilar. The key here for innovation is thinking thematically.

Which means?

Finding functional similarity in two or more taxonomically different things (Froehlich, Gibbert, & Hoegel 2014).

Apples and string cheese?

Better than that, an Apple chip and a Nike jogging shoe. Think about it this way. Many people enjoy listening to music while they exercise. Yet, it took more than a quarter of a century since the invention of the Walkman for the jogging shoe and the portable music industries to realize that they were, while dissimilar taxonomically speaking, similar thematically. Importantly, this similarity held commercial value. The product was the Nike+ which put an Apple iPod chip into a Nike shoe. The Nike+ uses the proprietary "Tune Your Run" technology, where the iPod displays information about the distance, pace, and calories burned. It even has a "PowerSong" button that lets the runner to select songs with a faster rhythm, enabling them to turn on, at the right moment, "the one song that gets you through the home stretch" (http://www.apple.com). *The Nike+ launch in 2006 was so successful that the company decided to make all its footwear iPod compatible by end 2008.*

I like this a lot. It gives the paired constraint model a role in *promoting* thematic and *precluding* taxonomic thinking.

Exactly, the idea is that we need a boundary to separate out promising opportunities in our "distant" search (Valikangas & Gibbert 2005). *"Distant" means taxonomically distant* **or** *dissimilar. But then, there are endless combinations/opportunities in a space that is simply not taxonomically similar. Most of these combinations will be random and senseless. What the paired constraint model does here is provide traction and guidance in this (theoretically) unlimited space (of taxonomically dissimilar opportunities) by precluding unpromising and pointing (promoting, as you call it) promising instances that are "somehow" (in my thinking, thematically) similar to the starting point.*

The Unique Contribution of the Paired Constraint Model

Innovation is a problem-solving process. Figure 8.1 represents the process as two loops, starting and ending at the domain.

The original domain (the white box) is the area of expertise whose *boundaries* (as in Cillier's model) define what it contains. The red box around the original domain represents the result of innovation—the domain *expands*. The domain provides the *basic expertise* in an individual's tool box. Connected to one tool box is the *borrowing* (paired constraints) that makes it bigger in a critical way. Our argument that *paired constraints* constitute a unique and critical borrowing is visualized by the two solution paths. *Variations*, paths that lead back to the original domain, are solutions that *meet current criteria.* They are both the product of and the problem with expertise in a domain. The problem is that experts get "stuck" in successful solutions. There are two reasons for this. One is that experts are rewarded (over and again) for solving the same sort of problem (over and again). The other is the result of that recurring reward. The most rewarded response becomes the default, the automatic solution path (with slight variations depending on situational specifics).

This is why paired constraints are critical. They ***preclude defaults***. The solution path with the red box representing *innovations*, by precluding variations on

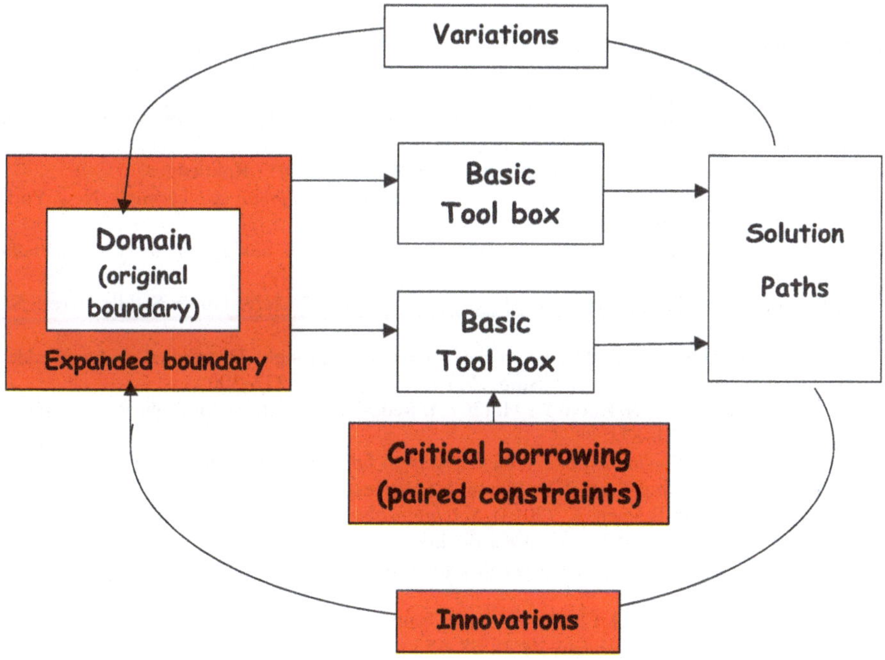

Fig. 8.1 Paired constraint process (shown in red)

dominant paths, promotes substitutions that *enlarge the boundaries* of the domain. This expansion is represented by the red box surrounding the original domain.

And since the system is an open loop, expanding the domain enlarges an expert's basic tool box, enabling further innovation and expansion into thematically similar domains. The key is that there are boundaries. The paired constraint model, in this sense, provides TWO kinds of actionable boundaries. It makes us aware of the boundary (or 'box') where we are, as you say, sometimes 'stuck, and at the same time alerts us to an expanded solution space in which we can roam freely (perhaps in thematically similar domains), thereby expanding our thinking.

I always like your expansions, Michael.

I always like your diagrams, but tell me why didn't you include the matrix?

That's simple Michael. The matrix categorizes innovation problems. The diagram shows how paired constraints uniquely enhance the solution process.

Uniquely and more usefully than the other models?

Yes, because, to paraphrase the title of a book—James Webb Young's *A Technique for Producing Ideas* (1940)—that was given to every junior copywriter at J. Walter Thompson. I like to think of the paired constraint model as a uniquely useful "technique for producing innovations."

Of course, the only way for our readers to judge is to put the "technique" into practice.

Next chapter.

References

Ahuja, G., & Katila, R. (2004). Where do resources come from? The role of idiosyncratic situations. *Strategic Management Journal, 25,* 887–907.

Cilliers, P. (1998). *Complexity and postmodernism.* New York, NY: Routledge.

Duggan, W. (2007). *Strategic intuition: The creative spark in human achievement.* New York, NY: Columbia Business School Publishing.

Eagleman, D., & Brandt, A. (2017). *The runaway species: How human creativity remakes the world.* New York, NY: Catapult.

Froehlich, J., Gibbert, M., & Hoegel, M. (2014). *Thematic thinking.* Upper Saddle River, NJ: Financial Times Press.

Garud, R., & Karnow, P. (2002). Bricolage versus breakthrough: Distributed and embedded agency in technology entrepreneurship. *Research Policy, 32,* 277–300.

Gupta, A., Hoopes, D. G., & Knott, A. M. (2015). Redesigning routines for replication. *Strategic Management, 36,* 851–871.

Isaacson, W. (2014). *The innovators: How a group of hackers, geeks, and geniuses created the digital revolution.* New York, NY: Simon & Schuster.

Liedtka, J., King, A., & Bennett, K. (2013). *Solving problems with design thinking: Ten stories of what works.* New York, NY: Columbia Business School Publishing.

Stokes, P. D. (2006). *Creativity from constraints: The psychology of breakthrough.* New York, NY: Springer.

Valikangas, L., & Gibbert, M. (2005). Boundary-setting strategies for escaping innovation traps. *MIT Sloan Management Review, 48*(3), 58–65.

Young, J. W. (1940). *A technique for producing ideas.* Lincolnwood, IL.: National Textbook Co.

Chapter 9
Putting Paired Constraints into Practice

How do you get to Carnegie Hall?
Practice, practice, practice.
Old joke

How do you get to innovate?
Practice, practice, practice.
Pat and Michael

Using any new tool takes practice. To help you put the paired constraint model into practice, there are three things we need to recap. The first: identifying the problem. The second: breaking it down into the structure of its solution. The third: making your tool box bigger. A recap is a reminder, somewhat re-said so that it will be read. In other words…

Identifying the Problem

This is not as simple as it sounds.[1] A few problems are given, most must be found, and many are missed.

Given Problems

The simplest are problems of ***necessity***. These are obviously given, somethings are missing and must be replaced (or in some cases, re-fined) in order for the problem to be resolved. The goal is the same, and the solution path is new.

---

[1]Or as it looks on the Matrix (Chap. 3).

P. D. Stokes and M. Gibbert, *Using Paired Constraints to Solve The Innovation Problem*, https://doi.org/10.1007/978-3-030-25771-2_9

Re-fining is interesting. Does that put the latest minor modification of the iPhone here?

I think so. The goal is always the same, to always be up-to-date, and never, never be out of the loop. That's why obsolescence is built into products like the iPhone. The new features are not as important as the fact that they are new. Novelty is necessary.

Partially given are problems of aspiration (what is next) and problems of opportunity (what is noticed). What is given in aspiration is an ***initial state***, a dominant style or solution the innovator aims (aspiration) to replace. What is given in opportunity is a ***glimpse of a goal state***, new and completely unspecified, but sufficient to start the solution-by-substitution process. Both problems—aspiration and opportunity—are difficult. They require new paths to new, not-yet-defined, goals. The difficulties are desirable. The solution paths expand (or even create) domains.

Pat wait, how does the glimpse of a goal start the substitution process?

Let me answer with my favorite example, the color wheel. What the wheel gave Monet was a glimpse of a completely new way to re-present the world. The first substitution pair became obvious: to show how light breaks up on things, preclude contrasts in value and promote contrasts in hue. There's something else the glimpse does. We've talked about things noticed because needed. The color wheel was different, it was not needed *until* it was noticed.

So, needed because noticed.

Something like that.

Found Problems

The difficulty with found problems is finding them.

Some are problems where replacement is ***not necessary***. The goal is unchanged. Existing solution paths reach the goal. Innovation depends on finding substitutions that can make an existing path more effective or more attractive. Netflix is a good example. It started with the same goal as Blockbuster (watching movies at home), but offered a path (on line, monthly fee, etc.) more effective and attractive than Blockbuster's (in store, late fees, etc.).

Pat, let me interrupt. This was indeed the case when Netflix started. What it became puts its subsequent innovations in the opportunity or aspirational categories.

Or maybe both.

Other found problems?

Many involve what you, Michael, call "***points of departure***." In its simplest form, a point of departure is part of an existing path that can be used to reach a new goal. The problem is found when the ***to-be-used part*** is noticed. For example, someone noticed that the path for the Pressman (a recording device) could easily morph—via a series of closely related substitutions—into one for the Walkman (a listening device).

Closely related like smaller ear phones being substituted for larger speakers.

Yes, before the smaller ear phones themselves morphed first into tiny ear buds and more recently into bigger (what I think are called) noise reducers. And Michael, I just thought of something else. This comes from my advertising experience. Often, problems are "found" for a product to solve.

Made-up problems?

Not completely, the problem has to have some relevance to a consumer's experience. What the advertising does is articulate the problem, make it salient.

An example, please.

Sure, there's an ad for Always in a recent issue of Living, one of Martha Stewart's magazines. Always is a sanitary pad. The headline announces the new problem: "60% of women are wearing the wrong size pad." The product (AlwaysMyFit™) solves the problem because it comes in five different sizes. It's hard to find an advantage in a parity product—Always just did. So, which category?

Problem found (wrong fit) on an existing path (single size). Next?

Next is practice using the model.

Wait, don't you think we should first point out where given and found problems are found on the Matrix?

Briefly.

Briefly, found problems belong to the transition categories (I and III). The transitions are from same to new path. Given problems, which need no transition from same to new path belong in Categories II and IV.

Solving the Innovation Problem Part I: Starting to Do Something New

To start, draw a problem space (like the one in Table 9.1).

Assuming that you've identified the initial state (what you are trying to resolve, refine, or replace) fill in the problem space, following this sequence:

- Enter the initial state and its criterion.
- Indicate the goal state, to whatever degree it can be specified.[2]
- Put all the elements/processes that define the initial state in the preclude column.
- Study the column carefully. Move necessary items to the retain column (Table 9.1).

[2]This will be simple if your problem is one of necessity. In this case, the goal is given. It will be difficult if your problem is one of aspiration or opportunity. You goal will only be fully defined when your solution path is fully specified.

Table 9.1 Problem space for ________________________

Initial state:			
Criterion:			
Search space:	**Constraint pairs**		
Retain	*Preclude*	→	*Promote*
Goal state:			
Criterion:			

You can now start the solution-by-substitution process.

- Pick one element to preclude. Move it to the top of the preclude column.
- Select a substitute. Be bold. Try the opposite of what you are precluding.
- Put it at the top of the promote column.

You have your first preclude-promote constraint pair.

- See where it leads. Remember, good ideas have "legs."
- If it goes nowhere, let it go.
- If it requires or suggests another pairing, keep it.
- If it is generative, it will set off a cascade of constraint pairs.
- Reiterate until your substitutions have constructed a solution path.

This probably will not be your final solution path. You will have to construct and consider multiple (possible/probable) solution paths. Some will be retained in part, and some precluded in total.

- Re-do. Start another problem space.
- Pick one element to preclude ….

This will take time. It is worth the time-taken. One path *will* prevail. You will not miss it.

Take the time. Practice, practice, practice.

Table 9.2 Problem space for ______________________

Initial state:			
Criterion:			
Search space:	**Constraint pairs**		
Retain	***Preclude***	→	***Promote***
Goal state:			
Criterion:			

Solving the Innovation Problem Part II: Continuing to Do Something New

To continue, *start over.*

Start by drawing a new problem space (like the one in Table. 9.2).

- Be like Monet.
- Make the goal state (from your first innovation) the initial state.
- Move the solution path (the promote column) from your first innovation into the preclude column.
- Study the column carefully. Move necessary items to the retain column.
- Continue as above (Table 9.2).

Reiterate whenever necessity, desirability, opportunity, or aspiration arise.

Making Your Tool Box Bigger

This recap is concise, but not inconsequential. Read carefully and remember.

- Substitutions and strategies for constructing solution paths are only found in your tool box.
- Borrowings make your tool box bigger.

- You should make it bigger because....
- The innovative-ness of your solutions depend *entirely* on its contents.

In other words, contents have consequences.

A critical, strategic content is the paired constraint model.

We hope this book has convinced you to borrow it and use it to solve your innovation problem.

Appendix A
Coming to Terms with Constraints

Constraints. In common usage, one-sided: limiting, restraining, precluding. In problem-solving usage, two-sided, pairs with different purposes: precluding something/promoting its substitute.

Criterion Constraints. Agreed upon standards that define a process, product, or style. The initial state defines the process, product, or style that the innovator wishes to recreate, refine, or replace. The goal state defines its recreation, refinement, or replacement, i.e., the innovation.

Paired Constraints. The *strategy* for constructing a solution path. One of each pair *precludes* something in the initial state; the other *promotes* a substitute. The sequence of substitutions is the solution path.

Problem Space. The *structure* for solving an innovation problem. The space has three parts. An initial state (the current situation), a goal state (the desired situation), and between the two, a search space in which paired constraints construct a solution path that changes the initial state into the goal state. The path is itself the innovation.

Problem Types

> ***Well-Structured Problems***. In a well-structured problem, the initial state, goal state, and search space are completely specified. Innovation is not possible in a well-structured problem.
>
> ***Ill-Structured Problems***. An ill-structured problem is not completely specified. This makes innovation possible.

Solution by Substitution. The *process* by which paired the constraints construct the solution path. One of each pair precludes something specific in the initial state. The other promotes a substitute. A productive pairing will require or suggest another, creating a substitution series. The promote column is the substitution series. It is also the solution path, e.g., the innovation.

P. D. Stokes and M. Gibbert, *Using Paired Constraints to Solve The Innovation Problem*, https://doi.org/10.1007/978-3-030-25771-2

The Solution Path. The promote column. The innovation.

Source Constraints. Specific to a domain or area of expertise. Things that an innovator works with or against.

Subject Constraints. Classifications. Into what category does an innovation belong?

Task Constraints. Materials and accepted methods for using them. Derived from source constraints. Found inside an expert's tool box.

The Tool Box. This is the one inside your head, the one you can't think outside of.

Basics. Content from the expert's domain. Basics come in two kinds: tools in your head and, by extension,[1] tools in your hands. The tools in your head constitute semantic memory (what you know); the ones in your hands, procedural memory (what you can do with what you know, i.e., skills).

Borrowings. Content from outside the expert's domain. Things noticed when needed.

---

[1]Via brain areas including basal ganglia, cerebellum, and motor cortex.

Appendix B
Opportunity Continued: Expanding *Only the NUMBERS Count*© in Second and Third Grades

This is where I (Pat) moved the expansion of my early math program (Chap. 7). The problem was expanding the initial solution path to satisfy the fluency criterion in higher grades. The opportunity was noticing how the Chinese multiplication table could be adapted to teach multiplication, division, and fractions as parts of an integrated pattern. Table A.1 shows the expanded problem space.

Criterion constraints. The initial state was current second- and third-grade curriculum. The goal was expanding the new curriculum. The goal criterion remained the same. The sub-goal (which stipulates the subject constraints) was mastery of multiplication, division, and unitary fractions.

Source and task constraints. My source was the Chinese multiplication table, morphed into a multi-operational chart for teaching multiplication and division simultaneously. Once the two operations were mastered, fractions were integrated into the multiplication/division patterns. The new multi-operation chart, shown in Fig. A.1, differs from our (American) multiplication table in *format, recitation,* and *application.*

Format. Notice that each vertical column begins with the square of its number. For example, the two's column starts with 4 (2 × 2) and continues (as ours does) in increments of 2. This is because 2 (1 × 1) is included in the one's column. As a result, there are 45 empty cells that help the children "see" the interconnected patterns. The one that is highlighted (at the intersection of the 2s column and the 3s row) is explained below.

Recitation. Recitation is similar to the Chinese in that each column is initially recited, not as an addition series (2, 4, 6, etc.), but as a multiplication series (two 2s are 4, two 3s are 6, two 4s are eight, etc.). It expands on the Chinese in that each multiplication problem is recited in two ways. Look at the highlighted intersection in Fig. A.1. Going down the two's column, a child would recite "two 3s are 6." Going across the three's row, she would also recite "three 2s are 6." When division is introduced the recitation would expand to include "6 divided by 2 is 3" and "6 divided by 3 is 2."

P. D. Stokes and M. Gibbert, *Using Paired Constraints to Solve The Innovation Problem*, https://doi.org/10.1007/978-3-030-25771-2

Table A.1 Problem space for second and third grades

Initial state:	Current curricula for second and third grades		
Search space	**Constraint pairs**		
	Preclude	→	**Promote**
	1. Standard multiplication table		Multi-operational table
	2. Recite table as addition		Recite as multiplication, division and fractions
	3. Separation of operations		Simultaneous learning of operations
Goal state: new curriculum (*Only The NUMBERS Count©*)			
Criterion: fluency: thinking in numbers, symbols, and patterns			
Sub-goal: linking patterns in multiplication, division, and fractions			

	1	2	3	4	5	6	7	8	9	10
1	1									
2	2	**4**								
3	3	6	**9**							
4	4	8	12	**16**						
5	5	10	15	20	**25**					
6	6	12	18	24	30	**36**				
7	7	14	21	28	35	42	**49**			
8	8	16	24	32	40	48	56	**64**		
9	9	18	27	36	45	54	63	72	**81**	
10	10	20	30	40	50	60	70	80	90	**100**

Figure A.1 Multi-operation chart used for multiplication, division, and fractions

Application. The uniqueness of the chart is that it is multi-operational. What the children are practicing are integrated patterns. Once they master the multiplication/division patterns, they learn that unitary fractions are a kind of division, for example, that one half (1/2) means divide by 2, and one-third (1/3) means divide by 3.[2] Just as in division, the answers (and the recitation) are at the

[2]1/3 is called both “one third” and “one of three equal parts.”

intersection highlighted in the figure: "one-half of 6 is 3, one-third of 6 is 2." Children now practice reciting, and also writing, the *entire pattern* (multiplication, division, fractions) for a given intersection. The written version for the intersection of 2 and 3 is:

$$\begin{array}{cc} 2 \times 3 = 6 & 3 \times 2 = 6 \\ 6 \div 2 = 3 & 6 \div 3 = 2 \\ 1/2\,\text{of}\,6 = 3 & 1/3\,\text{of}\,6 = 2 \end{array}$$

Third graders also added (3/8 + 2/8 = 5/8), decomposed (5/8 = 3/8 + 1/8 + 1/8), and multiplied (5/8 × 4/10 = 1/4) fractions, operations not usually introduced until fourth and fifth grades.

Results. Eighty-seven percent of the third graders taught with this curriculum from kindergarten through third grade scored at or above the proficient level on the PAARC math tests given in the Spring of 2017.

I wish I learned math this way.

Me too.

The manufacturer's authorised representative in the EU is Springer Nature Customer Service Centre GmbH, Europaplatz 3, 69115 Heidelberg, Germany. If you have any concerns regarding our products, please contact ProductSafety@springernature.com

Printed and bound by CPI Group (UK) Ltd, Croydon, CR0 4YY
30/09/2026
02225465-0005